Standard of Ministry of Water Resources of
the People's Republic of China

SL 197—2013
Replace SL 197—97

Specification for Survey of Water and Hydropower Projects

Drafted by:
Changjiang Institute of Survey, Planning and Design Research
Changjiang Spatial Information Technology Engineering Co., Ltd. Wuhan

Translated by:
Water Resources and Hydropower Planning and Design General Institute, MWR
China Water Resources Beifang Investigation, Design & Research Co., Ltd.
Changjiang Three Gorges Geotechnical Consultants Co., Ltd.
Yellow River Engineering Consulting Co., Ltd.
China Three Gorges University
Hebei Research Institute of Investigation and Design of Water Conservancy and Hydropower
Changjiang Spatial Information Technology Engineering Co., Ltd.

China Water & Power Press
Beijing 2017

图书在版编目（CIP）数据

水利水电工程测量规范：英文版：SL 197-2013 = Specification for Survey of Water and Hydropower Projects SL197-2013 / 中华人民共和国水利部发布. -- 北京：中国水利水电出版社，2017.7
ISBN 978-7-5170-5669-0

Ⅰ.①水… Ⅱ.①中… Ⅲ.①水利工程测量－技术规范－英文②水力发电工程－工程测量－技术规范－英文 Ⅳ.①TV221-65

中国版本图书馆CIP数据核字(2017)第181131号

书　名	Specification for Survey of Water and Hydropower Projects SL 197—2013
作　者	中华人民共和国水利部　发布
出版发行	中国水利水电出版社 （北京市海淀区玉渊潭南路1号D座　100038） 网址：www.waterpub.com.cn E-mail：sales@waterpub.com.cn 电话：（010）68367658（营销中心）
经　售	北京科水图书销售中心（零售） 电话：（010）88383994、63202643、68545874 全国各地新华书店和相关出版物销售网点
排　版	中国水利水电出版社微机排版中心
印　刷	北京瑞斯通印务发展有限公司
规　格	140mm×203mm　32开本　14.125印张　494千字
版　次	2017年7月第1版　2017年7月第1次印刷
定　价	**500.00元**

凡购买我社图书，如有缺页、倒页、脱页的，本社营销中心负责调换

版权所有·侵权必究

Introduction to English Version

China Water Conservancy and Hydropower Investigation and Design Association (CWHIDA) is a non-profit and nationalwide self-regulatory organization under the administration of the Ministry of Water Resources of the People's Republic of China. CWHID has members including Class A and major B companies and institutes in the China water conservancy and hydropower investigation and design industry. In order to satisfy the increasing market demand from international water conservancy and hydropower engineering construction, CWHIDA formulated *Translation Plan of Technical Standards 2015—2017* and organized some of its members to translate some investigation and design standards from Chinese to English.

Translation of the SL 197—2013 *Code for Surveying of Water Resources and Hydropower Engineering* was undertaken by the Water Resources and Hydropower Planning and Design General Institute, MWR, under the sponsorship of the CWHIDA.

This English version of this standard is identical in force to its Chinese original. In case of any discrepancy, the Chinese original shall prevail.

Translation of this standard is undertaken by the following organizations:

Water Resources and Hydropower Planning and Design General Institute, MWR

China Water Resources Beifang Investigation Design and Research Co., Ltd.

Changjiang Three Gorges Geotechnical Consultants Co., Ltd.

Yellow River Engineering Consulting Co., Ltd.

China Three Gorges University

Hebei Research Institute of Investigation and Design of Water Conservancy and Hydropower

Changjiang Spatial Information Technology Engineering Co., Ltd.

Translation task force includes SI Fu'an, XIE Jinping, LI Huizhong, HU Jie, ZHANG Guodong, WANG Haicheng, HE Bencai, LI Bingxing, YANG Peibing, ZENG Huaien, DUAN Shiwei, CUI Hongyan, CHEN Xiangyang, MING Tao, PANG Qingyan, YAN Jianguo, XIE Shiyu, YI Xingchen, ZHONG Hua, CHAI Zhiyong, LIU Yongbo and WANG Xuejiao.

This standard is reviewed by HUANG Shengxiang, CHEN Ting, SHEN Yunzhong, HAO Xiangyang, DING Wanqing, GUO Qiaozhen, QIN Wei, ZHAO Zhi, DING Xiaoli and LIU Jiahui.

China Water Conservancy and Hydropower Investigation and Design Association

Foreword

According to the plans for the formulation and revision of water industry technical standards of The Ministry of Water Resources (MWR), and the requirements of SL 1—2002 *Specification for the Drafting of Thechnical Standards of Water Resources*, the SL 197—97 *Code for Water Resources and Hydropower Engineering Surveying* (*in planning and design stage*) was revised in 2013, and its title was also changed into SL 197—2013 *Specification for Survey of Water and Hydropower Projects*.

This specification comprises 14 chapters and 7 annexes, and the technical contents include general provisions, terms and symbols, basic requirements, horizontal control survey, vertical control survey, digital topographic map surveying, aerial and space photogrammetry, terrestrial laser scanning and terrestrial photogrammetry, remote sensing interpretation, map compilation, special engineering surveys, GIS development, editing and warehousing of spatial data, acceptance of achievements and quality inspection and evaluation, etc.

The major revisions are as follows:

—Addition of terms and symbols;

—Deletion of sections corresponding to steel tap measurement, stadia ranging traverse, and traditional manual drafting;

—Emphasis of GNSS surveying in horizontal control survey and addition of RTK method;

—Addition of content on GNSS vertical control survey;

—Addition of content on the use of total stations, digital

levels, terrestrial laser scanning, etc. ;
　—Addition of content on digital mapping;
　—Addition of relevant provisions on special project surveys such as dike survey, shoreline utilization planning survey, water conveyance line survey, power transmission survey, water area survey, urban water works survey, the basin basic control survey, and regional ground subsidence monitoring;
　—Addition of provisions on remote sensing applications.

　The previous version SL 197—97 was replaced by the specification from 2013.

　The specification is approved by the Ministry of Water Resources of the People's Republic of China.

　The specification is interpreted by Water Resources and Hydropower Planning and Design General Institute, MWR.

　The specification is chiefly drafted by Changjiang Institute of Survey, Planning and Design Research, and Changjiang Spatial Information Technology Engineering Co, Ltd. Wuhan.

　This specification is jointly drafted by Yellow River Engineering Consulting Co. , Ltd. , China Water Resources Northeastern Investigation, Design & Research Co. , Ltd. , Hebei Research Institute of Investigation & Design of Water Conservancy and Hydropower, Jiangsu Engineering Exploration and Surveying Research Institute Co. , Ltd. , Xinjiang Water Resources and Hydropower Survey and Design Institute, Guangdong Hydropower Planning & Design Institute, Shenzhen Water Planning and Design Institute.

　This specification is published and distributed by China Water & Power Press.

　Chief drafters of this specification are YANG Aiming, YAN Jianguo, JIANG Benhai, DING Wanqing, GAO Zhiqiang,

WANG Haicheng, CHEN Yongqin, LI Yuping, XU Zhiqiu, ZHOU Chuansong, ZHAI Jianjun, CHEN Yuchang, ZHANG Li, WANG Xichun, GUO Zuojie, YE Qing, XU Yinglin, and XIONG Xun'an.

The technical responsible person for the specification review meeting is ZHANG Zhenglu.

The format examiner of this specification is WANG Qingming.

Contents

Introduction to English Version
Foreword
1 General Provisions ·· 1
2 Terms and Symbols ·· 5
   2.1 Terms ·· 5
   2.2 Symbols ··· 6
3 Basic Requirements ·· 11
4 Horizontal Control Survey ······································· 16
   4.1 General Requirements ····································· 16
   4.2 GNSS Survey ·· 18
   4.3 Triangular Network Survey ······························ 27
   4.4 Traverse Survey ··· 37
   4.5 Relevant Data Compilation and Submission ········· 43
5 Vertical Control Survey ·· 45
   5.1 General Provisions ·· 45
   5.2 Site Selection and Monumentation ···················· 47
   5.3 Leveling ·· 48
   5.4 Trigonometric Leveling with EDM ···················· 56
   5.5 GNSS Vertical Survey ···································· 64
   5.6 Relevant Data Compilation and Submission ········· 68
6 Digital Topographic Map Surveying ·························· 70
   6.1 General Provisions ·· 70
   6.2 Digital Mapping ··· 70
   6.3 Digital Topographic Map ································· 80
   6.4 Topographic Map Revision ······························ 83
   6.5 Topographic Map of Check ······························ 84

6.6	Document and Data Compilation	87
7	**Aerial and Space Photogrammetry**	**88**
7.1	General Requirements	88
7.2	Photo Control Point Design	105
7.3	Measurement of Photo Control Points	117
7.4	Annotation	118
7.5	Aerial Image Scanning	125
7.6	Digital Aerotriangulation	127
7.7	DLG Data Collection and Production of Map	136
7.8	Data Collection and Processing of Digital Elevation Model	147
7.9	Production of Digital Orthophoto Map	152
7.10	Airborne LiDAR Scanning Survey	156
7.11	Unmanned Air Vehicle (UAV) Low-altitude Digital Photogrammetry	164
7.12	Relevant Data Compilation and Submission	175
8	**Terrestrial Laser Scanning and Terrestrial Photogrammetry**	**178**
8.1	General Requirements	178
8.2	Terrestrial Laser Scanning Survey	179
8.3	Terrestrial Stereo (Multi-baseline) Photogrammetry	185
8.4	Relevant Data Compilation and Submission	189
9	**Remote Sensing Image Interpretation**	**191**
9.1	General Requirements	191
9.2	Preparation	192
9.3	Remote Sensing Image Processing	193
9.4	Establishment of Interpretation Signs	195
9.5	Remote Sensing Image Interpretation	196
9.6	Field Verification	196
9.7	Post Processing of Interpreted Information	198

9.8	Production of Thematic Map	198
9.9	Relevant Data Compilation and Submission	199

10 Map Compilation ... 201

10.1	General Requirement	201
10.2	Compilation of Topographic Map	201
10.3	Compilation of General Geographic Map	202
10.4	Compilation of Thematic Map	203
10.5	Atlas Compilation	206
10.6	Relevant Data Compilation and Submission	206

11 Special Engineering Survey ... 208

11.1	Engineering Survey of Land Acquisition and Resettlement	208
11.2	Levee Engineering Survey	220
11.3	Survey for Shoreline Utilization Planning	229
11.4	Water Conveyance Route Survey	233
11.5	Survey of Electric Power Transmission Line	240
11.6	Road Survey	248
11.7	Geological Exploration Survey	258
11.8	Survey in Water Area	269
11.9	Engineering Survey in Urban Water Projects	290
11.10	Basic Control Survey in River Basin	298
11.11	Monitoring of Regional Subsidence	301
11.12	Survey for Engineering Construction Control Network	304
11.13	Survey for Engineering Deformation Monitoring Network	316
11.14	Slope and Reservoir Bank Deformation Monitoring	325

12 GIS Development ... 333

12.1	General Requirement	333
12.2	Demands Analysis	335
12.3	Overall Design	338

12.4	Detailed Design	341
12.5	Software Coding and Testing	342
12.6	System Operation and Maintenance	344
12.7	Relevant Data Compilation and Submission	345

13　Editing and Warehousing of Spatial Data ··· 347

13.1	General Requirements	347
13.2	Digital Line Graphic	348
13.3	Digital Elevation Model	350
13.4	Digital Orthophoto Map	350
13.5	Digital Raster Graphic	351
13.6	Thematic Map Data	353
13.7	Relevant Data Compilation and Submission	353

14　Acceptance of Achievements and Quality Inspection and Evaluation ··· 355

14.1	Acceptance of Achievements	355
14.2	Quality Evaluation	358
14.3	Quality Inspection	362

Annex A　Horizontal Control Survey ··· 366
Annex B　Vertical Control Survey ··· 371
Annex C　Aerial Photogrammetry ··· 376
Annex D　Remote Sensing Image Interpretation ··· 388
Annex E　Special Engineering Survey ··· 391
Annex F　GIS Development ··· 401
Annex G　Product Acceptance and Quality Inspection and Evaluation ··· 413

1 General Provisions

1.0.1 This specification is hereby formulated to adapt to the current development level of surveying and mapping technologies, to unify the technical requirements on surveying technologies in water and hydropower projects, and to ensure that the quality of surveying and mapping products meets the needs of the construction projects.

1.0.2 The specification is applicable to surveying and mapping work in water and hydropower projects. In addition, construction survey in such projects should also be complied with the prevailing ministerial standard SL 52 *Technical Specifications for Construction Survey in Hydraulic and Hydroeletric Engineering*.

1.0.3 Before a survey, data shall be collected, site reconnaissance shall be made, and technical design shall be prepared. Quality control shall be done during a survey, and meanwhile a technical summary report shall be prepared after the completion of a survey. Technical design of a major project shall go through a design assessment while the results shall be evaluated.

1.0.4 Instruments and related equipment shall be tested and calibrated, and properly protected and maintained.

1.0.5 The software to be used shall be assessed and reviewed.

1.0.6 The collected surveying and mapping data shall be checked.

1.0.7 In this specification, root mean square error (RMSE) is used as a measure of accuracy and two times RMSE is regarded as the limit error.

1.0.8 The mapping scale shall be selected according to the actual needs of a construction project and the relevant provisions of the various design standards. For specific stage of design, the suitable topographic map for specific stage should be produced. When the next stage of the design needs to use the topographic map from the last stage, the topographic map shall be either revised or resurveyed.

1.0.9 This specification has referred to the following standards:

GB/T 7931　*Specifications for Aerophotogrammetric Field Work of* 1∶500　1∶1000　1∶2000 *Topographic Maps*

GB/T 12343.1　*Compilation Specifications for National Fundamental Scale Maps—Part* 1∶ *Compilation Specifications*

GB/T 12897　*Specifications for the First and Second Order Leveling*

GB/T 12898　*Specifications for the Third and Fourth Order Leveling*

GB/T 13923　*Specifications for Feature Classification and Codes of Fundamental Geographic Information*

GB/T 13977　*Specifications for Aerophotogrammetric Field Work of* 1∶5000　1∶10000 *Topographic maps*

GB/T 12979　*Specifications for Close-range Photogrammetry*

GB/T 13989　*Subdivision and Numbering for the National Primary Scale Topographic Maps*

GB/T 14511　*Specifications for Printing of Maps*

GB/T 14912　*Specifications for* 1∶500　1∶1000　1∶2000 *Field Digital Mapping*

GB/T 17798　*Geospatial Data Transfer Format*

GB/T 17942　*Specifications for National Triangulation*

GB/T 18314　*Specifications for Global Positioning System (GPS) Surveys*

GB/T 20257.1　*Cartographic Symbols for National Fundamental Scale Maps—Part 1： Specifications for Cartographic Symbols of* 1∶500, 1∶1000 & 1∶2000 *Topographic Maps*

GB/T 20257.2　*Cartographic Symbols for National Fundamental Scale Maps—Part 2： Specifications for Cartographic Symbols of* 1∶5000　1∶10000 *Topographic Maps*

GB/T 21010　Current and Use Classification

GB/T 24356　*Specifications for Quality Inspection and Acceptance of Surveying and Mapping Products*

GB/T 50138　*Standard for Stage Observation*

SL 52　*Code for Construction Survey of Hydraulic and Hydroelectric Engineering*

SDJ 336　*Technical Specifications for Concrete Dam Safety Monitoring*

CH/T 1007　*Metadata for Digital Products of Fundamental Geographic Information*

CH/T 1015.4　*Technical Rules for Producing Digital Products of* 1∶10000　1∶50000 *Fundamental Geographic Information—Part* 4∶ *Digital Raster Graphs*

CH/T 2008　*Specifications for Construction of the Continuously Operating Reference Stations Using Global Navigational Satellite System*

CH/T 8021　*Verification Regulation of Digital Aerial Photographic Camera*

CH/Z 3001　*Basic Requirements of Unmanned Air Vehicle Aerial Survey Safe Operation*

CH/Z 3002　*Technology Requirements of Unmanned Air Vehicle Aerial Photography System*

CJJ 61　*Technical Specification for Detecting and Surveying Underground Pipelines and Cables in City*

1.0.10 In addition to the provisions in this specification, the survey work in water and hydropower projects shall also comply with those in the other current relevant national standards.

2 Terms and Symbols

2.1 Terms

2.1.1 Related independent coordinate system

An independent coordinate system established by using a point in the survey area and whose coordinates are known in a national or local coordinate system as the starting point and the azimuth of the line from this point to another point in the national or local coordinate system as the starting azimuth, making no Gaussian projection correction to the side lengths, and taking the mean elevation surface of the survey area or of the buildings within the area, or a specified elevation surface as the projection plane for side lengths.

2.1.2 Boundary marker survey

A general term for marking out boundary lines in relocation of residents, land acquisition (for construction), treatment line of an urban area or of a special project, and a reservoir's normal water level line, etc.

2.1.3 Shoreline utilization planning survey

A surveying and mapping project in zones designated for protection or for rational development in an area between the waterfront control line and the outer control line of an inland river, lake, or estuary, etc.

2.1.4 Urban water works survey

A survey carried out for a city's water projects, water environment treatment projects, or water related construction projects, or for investigation and mapping of underground pipelines and cables buried underneath a water related construction site.

2.1.5 Unmanned air vehicle (UAV)

Short name for an unmanned aircraft, a reusable aircraft that is driven by a power system and operated by a wireless remote control or an onboard programmable control device.

2.1.6 Static draft

In water area surveying, the vertical distance between the transducer of a sounding instrument and the water surface when the surveying vessel is in a state of drifting or parking.

2.1.7 Pneumatic (dynamic) draft

In underwater surveying, the ship squat of the depth sounder transducer caused by the motion of the surveying vessel when it moves at its normal speed.

2.2 Symbols

a—Nominal constant error of a GNSS receiver or a distance measuring instrument;

B—Geodetic latitude;

b—Coefficient of nominal scale error of a GNSS receiver or a distance measuring instrument; average length of photo baseline;

C—Additive constant of a distance measuring instrument;

D—GNSS baseline length or average length; measured distance or side length;

D_0—Distance reduced to mean elevation plane in a survey area or to a specified elevation plane;

D_1—Distance reduced to a reference ellipsoidal surface;

D_2—Distance projected onto Gaussian plane;

D_{AB}—Horizontal distance;

$\overline{D}$—Mean horizontal distance of a measured side; horizontal distance between two stations;

d — Difference between forward and backward distances reduced to the same elevation plane; discrepancy between height differences;

d_i — Coordinate differences between common control points in different regional networks;

F — Length of a loop leveling line;

f — Correction to zenith distance for earth curvature and atmospheric refraction; focal length of an aerial camera;

f_β — Azimuth misclosure of a compound traverse or a loop traverse;

H — Relative flight height; depth of water; geodetic height;

h — Basic contour interval; tolerance of elevation difference between two ends of a measured side; elevation difference or height difference; depth of buried pipeline centerline;

h_0 — Known height difference (or known elevation difference);

H_m — Mean elevation of two ends of a measured side; height of geoid above the reference ellipsoidal surface in a survey area;

H_p — Mean elevation of a survey area or a specified elevation;

i_A, i_B — Heights of instrument at station A and station B respectively;

K, K_{AB}, K_{BA} — Coefficient of atmospheric refraction;

k — Ratio of denominator of photographic scale to that of map scale;

L—Length of a closed or loop leveling line; distance between two survey monuments; geodetic longitude;

l_A, l_B—Target heights;

L_X—Width of digital image in flight direction;

M—Denominator of a topographic map scale; denominator of a photograph scale;

m_0—Root mean square error of each distance measurement;

m_d—Root mean square error of the mean of reciprocal distance measurements;

m_D—Nominal accuracy of a DMEI;

m_H—Root mean square error of contour line elevation;

m_h—Root mean square error of the elevation of a densification control point;

m_g—Root mean square error of a fixed angle;

m_q—Root mean square error of unit weight parallax measurement;

m_s—Root mean square error of horizontal position of a densification control point;

m_{s1}, m_{s2}—Relative root mean square error of the length of start line;

$m_{\alpha 1}, m_{\alpha 2}$—Root mean square error of known azimuths at the ends of a connecting traverse;

M_Δ—Root mean square error per kilometer of mean elevation difference;

m_β—Root mean square error of angle observation;

m_C—Root mean square error of a check point;

m_N—Root mean square error of a common point;

N—Number of compound or loop traverses; number

of loop leveling lines;

n — Number of sides in a loop traverse; number of turning angles of a traverse; number of angles from an observation station; number of stations; number of reciprocal distance observations; number of height misclosures; number of triangles; number of points; number of baselines;

P — A priori weight;

P_i — Pixel size or image-scanning resolution;

P_x — Average amount of overlap between digital images in flight direction;

R — Multiplication constant of an EDM; radius of curvature of the Earth or mean radius of curvature of the Earth; length of a leveling line;

R_A — Radius of curvature of the normal section that contains the measured side;

R_m — Mean radius of curvature at the mid-point of a measured side on the reference ellipsoid;

R_{IS} — Resolution of image scanning;

R_O — Resolution of orthoimage;

S — Slope distance measurement; slope distance after meteorological, additive constant and multiplication constant corrections; cross-river distance in reciprocal leveling; size of a survey area; horizontal distance;

S_1, S_2 — Length of a starting side;

S_{AB} — Corrected slope distance between station A and station B;

T — Denominator of relative root mean square error of a distance;

W— GNSS loop network misclosure; angular miscolure of a triangle; miscloure of a leveling loop;

W_r—Tolerance of difference between observed and calculated angle;

W_s—Tolerance of azimuth misclosure of a closed traverse;

W_x, W_y, W_z—Misclosures of coordinate components of a loop GNSS network;

μ—Root mean square error of an unit weight observation;

y_m—Mean of abscissas;

Z, Z_{AB}, Z_{BA}—Zenith distance or zenith angle;

α—Ground slope angle; interior angles of a triangle;

β—Angle for distance transmission; interior angles of a triangle;

Δ—Miscloure between left and right angles of a traverse; station angular misclosure; misclosure between forward and backward height differences or between left and right route height differences; misclosure between adjustment results of two survey sessions;

ΔD_k—Additive constant and multiplication constant corrections;

Δh—Required height accuracy to be achieved;

Δy—Difference between horizontal coordinates of two endpoints of a measured side;

Δi—Misclosure between value measured in the field and value calculated for a check point;

ΔS—Tolerance of horizontal position error.

3 Basic Requirements

3.0.1 A current national coordinate system, or an independent coordinate system connected to a national coordinate system shall be adopted as the horizontal coordinate system, which is according to the mapping scale, selected as follows.

  1 For the topographic survey with a medium or small scale, the national coordinate system with 3° Gaussian conformal projection zone may be adopted. In a high altitude area, an independent horizontal coordinate system with the mean elevation plane in the survey area or a specified elevation plane as the projection plane.

  2 For the topographic survey with a large scale, the distance distortion due to length projection shall not be larger than 5cm/km.

  3 For the topographic survey with a large scale in an area with main structures located or at an important construction site, an independent horizontal coordinate system connected to a national coordinate system should be adopted where the mean elevation plane of the survey area or of the buildings should be used as the projection plane.

3.0.2 A current national height datum shall be adopted as the height reference system while the original height datum may be used continuously in a key flood control area of a river basin.

3.0.3 In remote areas where it is difficult to connect to the current national control points, independent horizontal and height coordinate systems may be adopted. In areas where there exist some horizontal and elevation control points, the existing control systems may be used continuously. The relationships between such systems and the current national systems shall be provided for coordinate transformation.

3.0.4 The same horizontal and height systems shall be used for different design stages of an engineering project.

3.0.5 Terrain classification and the various topographic map provisions shall comply with the following requirements:

    1 Terrain classification shall comply with the provisions given in Table 3.0.5-1.

Table 3.0.5-1 Terrain classification

Terrain type	Most areas in topographic sheet	
	Ground slope angle (°)	Elevation difference (m)
Plain	≤2	≤20
Hills	2-6	20-150
Mountains	6-25	—
High mountains	>25	—

    2 The basic contour interval of a topographic map should be selected according to Table 3.0.5-2.

Table 3.0.5-2 Basic contour intervals of topographic maps

Map scale	Basic contour interval (m)			
	Plain	Hills	Mountains	High mountains
1:500	0.5	0.5	1.0	1.0
1:1000	0.5 or 1.0	0.5 or 1.0	1.0	1.0 or 2.0
1:2000	0.5 or 1.0	1.0	1.0 or 2.0	2.0
1:5000	0.5 or 1.0	1.0 or 2.0	2.0 or 5.0	5.0
1:10000	0.5 or 1.0	1.0 or 2.0	5.0	5.0 or 10.0

    3 For topographic maps of 1:5000 and 1:10000 in scale, the map sheet division and their numbering shall comply with

GB/T 13989: *Subdivision and numbering for the national primary scale topographic maps*. The subdivision and numbering of large scale maps shall comply with the provisions in GB/T 20257.1 *Cartographic symbols for national fundamental scale maps - Part 1: Specifications for cartographic symbols of 1 : 500, 1 : 1000 & 1 : 2000 topographic maps*. For special topographic maps used in water and hydropower projects, square or rectangular map sheets should be used and numbered sequentially according to the survey areas.

4 The maximum root mean square errors of the horizontal position of a feature point relative to an adjacent control point shall comply with the provisions in Table 3.0.5-3.

Table 3.0.5-3 **Plane position RMSE of feature points on topographic map**

Map scale	Plain and hills (mm on map)	Mountains and high mountains (mm on map)
1 : 5000 - 1 : 10000	±0.5	±0.75
1 : 500 - 1 : 2000	±0.6	±0.8

Notes: 1. The permissible RMSE of the horizontal position of an underwater terrain point can be 2 times the values in the table.
2. The permissible RMSE of the horizontal position of a feature point in a covert and difficult area can be 1.5 times the values in the table, while the permissible RMSE in mountains and high mountains can be ±1.0mm on map.
3. The permissible RMSE of the horizontal positions of terrain points for topographic maps in special project surveys shall comply with the provisions in Notes 1 and 2 above.

5 The permissible elevation RMSE of contour lines shall comply with the provisions in Table 3.0.5-4.

6 The permissible elevation RMSE of a labeled terrain

point relative to any adjacent vertical control point shall comply with the provisions in Table 3.0.5-5.

Table 3.0.5-4　Elevation RMSE of contour lines

Terrain type	Plain	Hilly area	Mountainous region	High mountainous region
Elevation RMSE of contour lines	$\pm\frac{1}{3}h$	$\pm\frac{1}{2}h$	$\pm\frac{2}{3}h$	$\pm 1h$

Notes: 1. The h is the basic contour interval (m).
2. The elevation RMSE of the contour lines is calculated based on the difference between the elevations of evenly distributed check points in a map sheet and the corresponding elevations calculated from contour interpolation.
3. When 10m basic contour intervals are used, the permissible elevation RMSE of the contour lines shall be ±5m.
4. The permissible elevation RMSE of contour lines for covert and difficult areas may be 1.5 times the values in the table.
5. The permissible elevation RMSE of contour lines for areas under water may be 2 times the values in the table.

Table 3.0.5-5　Elevation accuracy of labeled terrain points

Terrain type	Plain and hills	Mountains and high mountains
1:5000-1:10000	$\pm\frac{1}{4}h$	$\pm\frac{1}{3}h$

Notes: 1. The h is the basic contour interval (m).
2. When a 10m basic contour interval is adopted for a mountainous region or a high mountainous region, the permissible elevation RMSE shall comply with the provisions defined for 5m basic contour interval.

3.0.6　The density of the labeled terrain points and the number of decimal places of the labels shall comply with the following provisions.

1　10-20 points per 100cm^2 should be noted on a map for plains and hills.

2 8 – 15 points per 100cm^2 should be noted on a map for mountains and high mountains.

3 If the contour interval is 0.5m, the elevation note shall be 0.01m. Otherwise, it shall be 0.1m.

3.0.7 For layering of data in digital topographic products and the naming of the layers, refer to provisions in GB/T 14912.

3.0.8 Electronic notebook or data terminal should be used for recording field observations.

4 Horizontal Control Survey

4.1 General Requirements

4.1.1 Horizontal control survey may be divided into the basic horizontal control survey, topographic horizontal control and station horizontal control, etc. It may adopt GNSS (network) survey, trigonometric (network) survey and traverse (network) survey etc.

4.1.2 The basic horizontal control may be divided into the second, third, fourth, fifth class. All of them may be used as the primary control of survey area, and the order arrangement and accuracy requirements shall meet the requirements specified in Table 4.1.2.

Table 4.1.2 The order arrangement and accuracy requirements of horizontal control

Horizontal control order	Map Scales		Accuracy requirement on the map (mm)
	1:500	1:1000 1:2000 1:5000 1:10000	
Basic horizontal control	second, third, fourth, fifth class		Allowable positioning mean square error of weakest adjacent point is ±0.05
Topograpic horizontal control	first order	first order ↓ second order	Allowable positioning mean square error of the bottom level mapping control point to adjacent basic horizontal control point is ±0.1
Station horizontal control	The station	The station	Allowable positioning mean square error of the station to the adjacent mapping control point is ±0.2

Notes: 1. when map scale 1:500 is adopted, allowable positioning mean square error of weakest adjacent point is ±5 cm for the second, third, fourth, fifth basic horizontal controls.
2. when conditions are favorable, stations can be directly densified on the basis of basic horizontal control. For small area, topograpic horizontal control can be used as the primary control.
3. under the premise that the accuracy requirement is satisfied, control network can be set order by order or skipping orders.

4.1.3 The basic horizontal control points shall be marked through burying permanent monument, and a draft should be drawn to record the position of the station. The specifications for the buried monument shall comply with Annex A.

4.1.4 The density of topographic horizontal control points by total station survey should meet the needs of topographic survey, and shall not be less than the specifications in Table 4.1.4.

Table 4.1.4 Density of topographic horizontal control points

Map scale	1:500	1:1000	1:2000	1:5000	1:10000
Number of topograpic horizontal control points (per km^2)	32	12	4	2	1

4.1.5 Significant digit in office computation of horizontal control network shall be determined in accordance with regulations of Table 4.1.5.

Table 4.1.5 Regulations of significant digit in office computation of horizontal control network

Order	Observation value of direction and the correction (″)	Observation value of side length and the correction (m)	Side length and coordinate (m)	Azimuth (″)
Second-fourth class	0.1	0.001	0.001	0.1
Fifth order	1	0.001	0.001	1
Topographic control	1	0.001	0.01	1

4.2 GNSS Survey

4.2.1 The control network for GNSS survey can be divided into five classes. The distance between adjacent points and accuracy requirement of each order should be implemented according to the regulation of Table 4.2.1.

Table 4.2.1 Accuracy classification of GNSS control network and the regulation of distance among adjacent points

Order	Average distance among adjacent points (km)	Fixed error a (mm)	Proportional error b (mm/km)	The length relative mean square error of weakest adjacent points
Second class	8 – 13	≤10	≤2	1/150000
Third class	4 – 8	≤10	≤5	1/80000
Fourth class	2 – 4	≤10	≤10	1/40000
Fifth class	0.5 – 2	≤10	≤20	1/20000
Topograpic horizontal control	0.2 – 1	≤10	≤20	1/4000

4.2.2 GNSS network design should meet the following requirements:

1 The GNSS networks layout of all classes can be polygon or closed line, in which the minimum distance between adjacent points should not be less than 1/3 of the average distance between adjacent points, and the maximum distance should not be longer than 3 times the average distance between adjacent points.

2 When connecting survey of new GNSS network and the original control network is employed, the survey points in the original control network should not be less than 3, and the connecting survey points should be evenly distributed in the new network. In the area where densified control network is required by conventional measurement methods, pairs of GNSS points should be designed and be mutually visible.

3 When the baseline length is larger than 20km, the static observation should be adopted with time span of class C GNSS network in GB/T 18314.

4 The second, third, fourth class GNSS control networks shall be adopted with the layout of network-connecting type or edge-connecting type. The fifth-order or topographic control GNSS network may be adopted with the layout of point-connecting type.

5 The number of independent baselines of a closed or connecting traverse in a GNSS control network shall meet the requirements of Table 4.2.2.

Table 4.2.2 **The regulation on independent baseline number of closed or connecting traverse in a GNSS control network**

Survey class	Second class	Third class	Fourth class	Fifth class	Topographic horizontal control
Line number of close-loop or connecting traverse	⩽6	⩽8	⩽8	⩽10	⩽10

4.2.3 The location of GNSS point should be open overhead, and the elevation angle of obstacles should be not more than 15°. It shall be far away from large water area, high power transmitters or high voltage power line, and the distance should not be less than 50m.

4.2.4 The main technical requirements of each order of GNSS horizontal control survey should meet the regulations of Table

4.2.4-1 to Table 4.2.4-3.

Table 4.2.4-1 The basic technical requirements of GNSS static survey

Order		Second class	Third class	Fourth class	Fifth class	Topograpic horizontal control
The GNSS receiver type		Double frequency	Double frequency or single-frequency	Double frequency or single-frequency	Double frequency or single-frequency	Double frequency or single-frequency
Static	Satellite elevation angle	≥15	≥15	≥15	≥15	≥15
	The number of observation session	≥2	≥1.6	≥1.4	≥1.2	≥1
	The length of observation session (min)	≥90	≥60	≥45	≥30	≥15
	The number of efficient observation satellites	≥5	≥4	≥4	≥4	≥4
	Data sample interval (s)	10-30	10-30	10-30	10-30	10-30
	PDOP	≤6	≤6	≤6	≤8	≤10

Note: The number of observation session is no less than 1.6, 1.4 or 1.2, which means when network observation model is adopted, each station should be observed for at least one observation session, and the station mounted twice shall not be less than 60%, 40% or 20% of the total number of GNSS network station.

Table 4.2.4-2 The main technical requirements of RTK survey of fifth-order horizontal control points

Class	Average length among adjacent points (m)	RMSE of location (cm)	Relative mean square error of side length	Distance from base station (km)	Observation times	Order of reference point
Fifth (first)	1000	≤5	1/20000	≤5	≥4	Fourth class or higher
Fifth (second)	500	≤5	1/10000	≤5	≥3	Fifth class or higher

Notes: 1. RMSE of location means the error of the control points relative to nearest datum point.
2. When single base station is adopted, it should be replaced at least once.
3. When Network RTK control survey is adopted, the distance between rover station and base station may not be restricted, but rover station shall be within the effective range of the network.

Table 4.2.4-3 The main technical requirements of RTK survey of topographic horizontal control points

Distance from base station (km)	Observation times	Order of starting point
≤5	≥2	Fifth order or higher

Note: When network RTK control survey is adopted, the distance between rover station and base station may not be restricted, but rover station shall be within the effective range of the network.

4.2.5 When GNSS survey is implemented, it should meet the following requirements:

1 The allowable centering error is 2mm when the GNSS antenna is mounted, and the antenna height should be measured with the accuracy of up to 1mm.

2 During GNSS observation, use of radio communication equipment near the GNSS antenna is not allowed.

3 In the GNSS survey, necessary information shall be recorded correctly such as the name and number of point, receiver device model and serial number, antenna height, observation time and so on.

4.2.6 The relative positioning results of GNSS survey shall meet the following requirements:

1 In baseline computation, the elevation mask angle should be set to 20°.

2 For the reference point used for baseline computation, its observation time for single point positioning should not be less than 30min.

3 During the data processing, the horizontal and vertical offsets between the antenna phase center and marker of station shall be taken into consideration.

4 The deleted synchronization phase observation data should not be more than 10% of total observation data in baseline solution.

5 Length difference of repetition baseline measurement shall meet the requirements of Equation (4.2.6-1):

$$d_s \leqslant 2\sqrt{2}\sigma \qquad (4.2.6-1)$$

$$\sigma = \sqrt{a^2 + (bD)^2} \qquad (4.2.6-2)$$

Where:

d_s = the length difference of repetition baseline measurement, mm;

σ = the MSE of length of corresponding baseline, mm;

a = fixed error of the corresponding control network, mm;

b = proportional error coefficient of the corresponding control network, mm/km;

D = the average side length, km.

6 Closure error of each coordinate component of independent baseline loop and closure error of the overall length of loop wire shall meet the requirements of Equation (4.2.6-3):

$$\left.\begin{array}{l} W_x \leqslant 3\sqrt{n}\,\sigma \\ W_y \leqslant 3\sqrt{n}\,\sigma \\ W_z \leqslant 3\sqrt{n}\,\sigma \\ W = \sqrt{W_x^2 + W_y^2 + W_z^2} \\ W \leqslant 3\sqrt{3n}\,\sigma \end{array}\right\} \quad (4.2.6-3)$$

Where:

n = the number of independent baselines in the loop;

W = closure error of overall length of independent baseline loop wire, mm.

7 Allowable closure error of each coordinate component of dependent baseline loop is half of that of independent baseline loop.

4.2.7 If the observation data is not adequate or the closure error fails to meet the regulations of Article 4.2.6, the related baselines or synchronizing patterns shall be resurveyed.

4.2.8 Unconstraint adjustment of GNSS network shall be in accordance with the following regulations:

1 Three-dimensional (3D) unconstraint adjustment shall proceed in WGS-84 coordinate system, and provide the 3D coordinates of each observed point in WGS-84 coordinate system, and the corrections, baseline lengths, baseline azimuths and the related accuracy information, etc. of the three observed coordinate difference value of each baseline vector.

2 Absolute value of the baseline vector correction of unconstrained adjustment should not exceed 3 times the allowable mean square error of baseline length of corresponding orders.

4.2.9 Constrained adjustment of GNSS network shall meet the following regulations:

1 Two or three dimensional constraint adjustments shall proceed in the national coordinate system or local coordinate system.

2 The known coordinates, distances or azimuths may be used as constraints or weighted constraints.

3 Adjustment results of observed points shall be output with the two or three dimensional coordinates in the corresponding coordinate system, the corrections of baseline vector, baseline length, baseline azimuth and related accuracy information. If needed, the coordinate transformation parameters and their accuracy information shall also be output.

4 The relative mean squared error of weakest side of control network's constraint adjustment shall meet the regulations of corresponding order in Table 4.2.1.

4.2.10 Horizontal control survey adopting the RTK method shall meet the following requirements:

1 Coordinate transformation parameters can directly employ the result of GNSS network's constrained adjustment, or compute the transformation parameters with 4 or more matching points evenly distributed in the peripheral and central surveyed area; when the surveyed area is large and needs partition to determine the coordinate transformation parameters, the matching points of adjacent partition areas should not be less than 2 points.

2 The reference station should meet the following regulations:

　　1) When the network RTK is adopted, the establishment of reference station network points should comply

with the requirements of CH/T 2008.
2) The long-time and frequent used self-established reference stations should adopt the observation post with forced centering device.
3) The self-established reference stations should be set up at higher order control points.
4) If radio link is used to transmit data, the reference station should be set at relatively higher position in the survey area. If mobile communication is used to transmit data, the reference stations shall be set at the position with mobile communication signals in the survey area.
5) When the radio link communication is selected, the data chain shall be set to arrange working frequency.
6) The instrument type, radio type, radio frequency, antenna type, data port, bluetooth port and so on should be set up in accordance with the instrument software.
7) Parameters such as the coordinates of reference station, data unit, scale factor, projection parameters and antenna height of receiver, shall be set up.

3 The rover station shall comply with the following regulations:
1) The rover station of network RTK shall obtain the authorization of the system service, and maintain the data communication with the service control center in the effective service area.
2) The transformation parameters between two coordinate systems shall be set up using the data collector of rover station, and shall be set up in the same way.

3) RTK rover station should not be used to collect the observations in the hidden zone, the large water area or near the strong sunshine and power interference source.
4) Before observation, the instrument shall be initialized and the fixed solution can be obtained. The communication chain should be disconnected and then the instrument should be reinitialized when the fixed solution is not available for a long time.
5) For each time of observation, the rover station shall be reinitialized.
6) Centering and leveling shall be ensured with tripods for the rover station, when the fifth class or topographic control RTK is adopted. Epoch number of each observation should not be less than 20 and the sampling interval should be 2 – 5s. The tolerant differences among each surveyed horizontal coordinate is 4cm for fifth class, and 0.1mm in the map for topographic control.
7) During operation, if satellite signal is loss of lock, it shall be reinitialized, and only after relocation is qualified with the coincidence point check can the operation go on.
8) Before each operation or after the re-erection of the reference stations, the new operation shall be checked with at least one known point of the same order or higher order, and the allowable coordinate difference is 7cm.
9) For the fifth class RTK survey, the allowable residual of transformed coordinates is 2cm, and for topographic control RTK allowable residual of transformed coor-

dinates is 0.07mm in the map.

10) The horizontal Permissible convergent accuracy of single observation of control points is set to 2cm by the data collector.

11) When post-processing kinematic survey is operated, the rover station shall obtain a fixed solution after the static observation of 10 – 15min, and then carry out kinematic survey under the premise of not losing the initiation state.

4 The collected data of RTK field work shall be backed up and checked in time with both field work and office work.

4.3 Triangular Network Survey

4.3.1 Triangular network survey can be divided into four orders according to accuracy; the main technical requirements of each order shall meet the regulations of Table 4.3.1.

Table 4.3.1 Main technical requirements of triangular network survey

Class	Angulation mean squared error (″)	Triangular maximal closure error (″)	Average side length (km)	Number of rounds of horizontal angle observation			Relative RMSE of distance measurement	Length relative RMSE of the weakest side
				DJ1	DJ2	DJ6		
Second class	±1.0	3.5	8 – 13	9	—	—	1/300000	1/150000
Third class	±1.8	7	4 – 8	6	9	—	1/160000	1/80000
Fourth class	±2.5	9	2 – 4	4	6	—	1/120000	1/40000
Fifth class	±5.0	15	1 – 2	1	3	6	1/60000	1/20000
	±10.0	30	0.5 – 1	—	2	4	1/30000	1/10000

4.3.2 Site selection of the control point of triangular network shall meet the following requirements:

 1 Adjacent sites should be well intervisible, and the distance from line of sight to the impediment should not be less than 2m for the second order, and not be less than 1.2m for the third and fourth class, and should be in accordance with the principle that ensures the convenience of observation, and not be affected by the lateral refraction under the fourth class.

 2 The measured side should be selected on the location with the similar ground coverages, and should not be selected above the heating elements like chimney, heat emission tower, heat emission pool. There should be no impediments like branches, power lines on the measured line. The measured line shall be more than 1.3m far from the impediments and avoid the interference of strong electromagnetic fields such as high voltage power lines, and the objects with reflective surface behind sight line shall be avoided.

 3 The inclination of measured side should not be oversized. When bilateral trigonometrical leveling is adopted to measure the height difference, the height difference shall be less than the permissible tolerance computed by Equation (4.3.2).

$$h \leqslant \frac{20D}{T} \times 10^3 \qquad (4.3.2)$$

Where:

 h = the height difference of the two end points of measured side, m;

 D = the measured side length, m;

 T = the denominator value of relative mean squared error of the measured side requirement.

4.3.3 Field observation of triangular network shall meet the following regulations:

1 Total station should be adopted to observe the horizontal directions, distances and zenith angles.

2 Prism used in range measurement shall match the total station, and be consistent with calibration of instrument.

3 Before observation, the temperature of instrument shall be consistent with the ambient temperature. During observation, bubble center position should not deviate from the compensation scope of compensator. When the bubble position is close to the permissible tolerance of deviation, the instrument shall be re-levelled after current observation round is finished.

4.3.4 Observation of horizontal angle of triangular network shall be in accordance with the following regulations:

1 Observation of horizontal direction should adopt the direction observation method, and the operation procedure of one round and grouping observation are implemented according to the relevant regulations of GB/T 17942.

2 The technical requirements of direction observation method shall be in accordance with the requirements of Table 4.3.4.

Table 4.3.4 The technical requirements of direction observation method of horizontal angle   Unit: (″)

Class	Instrument type	Difference of two readings	Misclosure of round in half round	Difference of 2C deviation in one round	Difference of each round of one direction
Fourth class or above	DJ1	1	6	9	6
	DJ2	3	9	13	9
Fifth class or above	DJ2	3	12	18	12
	DJ6	—	24	—	24

Note: When the vertical angle in the observation direction exceeds ±3°, difference of 2C deviation in this direction may be compared according to adjacent rounds in the same direction.

3 The supplementary survey or resurvey of the observation of horizontal angles shall be implemented in the following cases:

  1) When the misclosure of round exceeds the limit or difference of 2C deviation of initial direction exceeds the limit, the round shall be reobserved.

  2) When difference of 2C deviation or difference of each round of one direction exceeds the limit, it is allowed to reobserve the only direction out of limit.

  3) In one complete round (including circle-left and circle-right rounds), if the number of reobserved directions exceeds 1/3 of the total number of directions, the round shall be reobserved. When the number of reobserved complete rounds exceeds the number of total complete rounds, this station shall be reobserved.

4.3.5 The classification of distance measuring instruments and demarcation of accuracy orders should be implemented according to the following regulations:

  1 The distance measuring instruments are divided into four orders according to the factory nominal accuracies, see Table 4.3.5.

Table 4.3.5 **Technical requirements of range finder level**

Accuracy class	Absolute value of nominal accuracy per km ranging (mm)
I	$\|m_D\| \leqslant 2$
II	$2 < \|m_D\| \leqslant 5$
III	$5 < \|m_D\| \leqslant 10$
IV (Substandard order)	$\|m_D\| > 10$

  2 Nominal accuracy of distance measuring instrument shall be expressed according to Equation (4.3.5).

$$m_D = \pm (a + bD) \tag{4.3.5}$$

Where:

a = fixed error of nominal accuracy, mm;

b = proportional error coefficient of nominal accuracy, mm/km;

D = measured length, km.

4.3.6 Range observation of triangular network shall meet the following regulations:

1 The allowable centering error of survey station and reflector station is 2mm.

2 Before operation of range observation, the instrument shall be correctly settled, and the height of instrument and height of prism should be measured with mm accuracy. When the ranges of the triangular network of the fourth class or above are measured, the height of instrument and height of prism respectively shall be measured once before and after range observation, and adopt the average value.

3 The range survey should be carried out during the best observation period of a day. When it is carried out on a sunny day, umbrella shall be setup to shade the survey instrument and meteorological instrument.

4 Observations of meteorological data shall meet the regulations of the Table 4.3.6.

Table 4.3.6 Requirements of observation of meteorological data

Class	The minimal reading		Time interval of determination	The use of the meteorological data
	Temperature (℃)	Atmospheric pressure (Pa)		
Fourth class and above	0.2	50	Determine once at the beginning and end of the each side observation respectively	The average value of two ends of each side
Fifth class and below	0.5	100	Determine once for each side observation	The data on the endpoint of survey station

4.3.7 Main technical requirements for the range observation of triangular network should satisfy the regulations of the Table 4.3.7.

Table 4.3.7 Main technical requirements of range

Class	Order of distance measurement instrument	number of round of each side	Reading difference limit of one round (mm)	Difference limit among rounds (mm)	Difference limit of forward and backward observations
Second class	I	Four rounds forward and backward respectively	2	3	$2(a+bD)$
	II	Four rounds forward and backward respectively	5	7	
Third class	II	Four rounds forward and backward respectively	5	7	
	III	Four rounds forward and backward respectively	10	15	
Fourth class	II	Four rounds forward and backward respectively	5	7	
	III	Four rounds forward and backward respectively	10	15	
Fifth class	III	Three rounds on one direction	10	15	—
	IV	Four rounds on one direction	20	30	
Topograpic control	IV	Two rounds on one direction	20	30	

Notes: 1. one round means instrument sights target one time and read value four times.
2. forward and backward observations may also be replaced by measurement of different period of time instead.
3. Difference of forward and backward observations should be obtained only when the slant distances are reduced to horizontal distance on the same plane.

4.3.8 Resurvey, adoption and neglect of range observation achievements shall be in accordance with the following regulations:

1 Observation achievements with the errors larger than permissible tolerances of Table 4.3.7 shall be remeasured.

2 When reading differences exceed the permissible tolerances in one complete round, one additional reading may be measured again. When additional reading also exceeds the permissible tolerance, the total rounds shall be measured.

3 When differences among different complete rounds exceed the limit, two rounds may be reobserved and big or small deviation shall be excluded, and then the average value is adopted if the examination of limit passes. When reobservation exceeds the limit, the total survey station shall be resurveyed.

4 If the discrepancy of forward and backward observations or discrepancy of different periods of observations exceeds the permissible tolerance, the reason shall be analyzed and distance single direction shall be resurveyed. If resurvey exceeds the limit, the distances of forward and backward observations shall be resurveyed.

4.3.9 Correction for the range observation of slope side may be calculated using the elevation difference from leveling or trigonometrical leveling with electromagnetic distance measurement. When the elevation difference is determined by trigonometrical leveling with electromagnetic distance measurement, bilateral observations shall be carried out. Height difference δ_h of forward and backward observations shall meet the requirement of Equation (4.3.9).

$$\delta_h \leqslant 0.1S \times 10^{-3} \qquad (4.3.9)$$

Where:

δ_h = height difference of forward and backward observations, m;

S = the measured distance, m.

4.3.10 Technical requirements of zenith angle measurements of bilateral observations shall be in accordance with the relevant regulations of Section 5.4. Mean squared error of zenith angle measurements of bilateral observations shall meet the requirements of Equation (4.3.10).

$$m_z \leqslant \frac{\sqrt{2}\rho}{5T\cos Z} \qquad (4.3.10)$$

Where:

m_z = mean squared error of zenith angle measurement, ($''$);

T = the denominator of relative mean squared error of requirement of range observation;

Z = zenith angle, (°);

ρ = the adopted value is 206265, ($''$).

4.3.11 The correction and reduction of range observation shall be in accordance with the following regulations:

1 The slope distance of range observation shall have meteorological correction, instrument addition and multiplication constant correction, and then horizontal distance can be calculated. If slope distance needs the addition of the correction of precision measurement frequency, this correction shall be added before the calculation of addition constant and multiplication constant.

2 Meteorological correction of range observation shall be calculated with the formulae given by the specification of total station or distance measurement instrument.

3 Addition and multiplication constant corrections of range observation shall be calculated with Equation (4.3.11-1).

$$\Delta D_k = RS + C \qquad (4.3.11-1)$$

Where:

$\Delta D_k =$ correction of addition and multiplication constant, mm;

$R =$ multiplication constant obtained by the verification of distance measurement instrument, mm/km;

$C =$ addition constant obtained by the verification of distance measurement instrument, mm;

$S =$ slope range observation, km.

4 If the reduction of range observation is horizontal distance, the calculation shall be carried out according to the following requirements:

1) When trigonometric leveling is adopted, horizontal distance is calculated with Equation (4.3.11-2).

$$D = S\sin(Z - f) \qquad (4.3.11-2)$$

$$f = (1 - K)\frac{S}{2R}\rho \qquad (4.3.11-3)$$

Where:

$D =$ horizontal distance of range observation, m;

$S =$ the slope distance of range observation after meteorological correction, instrument addition and multiplication constant correction, m;

$Z =$ zenith angle measurement, (°);

$f =$ the correction of zenith angle by considering earth curvature and atmospheric refraction, (″);

$K =$ the local atmospheric refraction coefficient;

$R =$ radius of curvature of the Earth in measurement area, m.

2) Using the elevation difference, the horizontal distance is calculated with Equation (4.3.11-4).

$$D = \sqrt{S^2 - h^2} \qquad (4.3.11-4)$$

Where:

    $h =$ elevation difference of the two endpoints of observed side, m.

5 Horizontal distance of observed side reduced to the average or given elevation plane of surveyed area shall be calculated with Equation (4.3.11-5).

$$D_0 = D\left(1 + \frac{H_p - H_m}{R_A}\right) \quad (4.3.11-5)$$

Where:

    $D_0 =$ distance reduced to the average or given elevation plane of surveyed area, m;

    $H_p =$ the average or given elevation of surveyed area, m;

    $H_m =$ average height of two endpoints of observed side, m;

    $R_A =$ radius of curvature of the normal section line of observed side, m.

6 Horizontal distance of observed side reduced to the surface of reference ellipsoid shall be calculated with Equation (4.3.11-6).

$$D_1 = D\left(1 - \frac{H_m + h_m}{R_A + H_m + h_m}\right) \quad (4.3.11-6)$$

Where:

    $D_1 =$ distance reduced to the surface of reference ellipsoid, m;

    $h_m =$ elevation difference of geoid above reference ellipsoid in surveyed area, m.

7 The distance reduced from the surface of reference ellipsoid to the Gauss projection plane is calculated with Equation (4.3.11-7).

$$D_2 = D_1\left(1 + \frac{y_m^2}{2R_m^2} + \frac{\Delta y^2}{24R_m^2}\right) \quad (4.3.11-7)$$

Where:

D_2 = side distance on the Gauss plane, m;

y_m = abscissa average value of the two endpoints of observed side, m;

Δy = abscissa increment of the two endpoints of observed side, m;

R_m = mean radius of curvature of the side midpoint on the reference ellipsoid surface, m.

4.3.12 Mean squared error of angle observations in the triangular network shall be computed according to Equation (4.3.12).

$$m_\beta = \pm \sqrt{\frac{[WW]}{3n}} \quad (4.3.12)$$

Where:

W = triangular misclosure, (″);

n = the number of triangles.

4.3.13 During the adjustment of triangulation network, the angle or direction observations and range observations shall be regarded as the observations involved in the adjustment. The adjustment shall adopt the strict adjustment method. Accuracy evaluation after adjustment shall include RMSE of surveyed angles, relative RMSE of side length, RMSE of point position, point position error ellipsoid parameter, etc.

4.4 Traverse Survey

4.4.1 The main technical requirements of traverse survey shall meet the following regulations:

  1 Main technical requirements for the traverse survey of the third, fourth, fifth class shall be in accordance with the requirements of Table 4.4.1-1.

  2 Main technical requirements of topographic control traverse survey shall be in accordance with the regulations of Table 4.4.1-2.

3 In the traverse network, the length from higher order point to node and from node to node should not be greater than 0.7 times the specified length of corresponding order in Table 4.4.1-1 and Table 4.4.1-2. The length of higher order point to higher order point should not be greater than 1.5 times the specified length of corresponding order in Table 4.4.1-1 and Table 4.4.1-2.

4.4.2 Layout of traverse shall meet the following requirements:

1 Each order traverse should be laid in the shape of straight line.

2 When the traverse control is used as intensified network, the single connecting traverse may be adopted. And the annexed traverse should be set up in the straight line shape and the lengths of each side should be approximately equal.

3 Length of single orientation angle topographic control traverse with electro-optical distance measurement shall be 0.8 times the length of equal order annexed traverse. Length of no orientation angle topographic control traverse with electro-optical distance measurement shall be 0.6 times the length of equal order annexed traverse.

4 No orientation angle topographic control traverse with electro-optical distance measurement should constitute traverse network. When it is a single traverse with no orientation angle, there shall be the check of known points. When the known points are advanced points, the difference of point position or transverse displacement shall not be greater than 0.2mm on the map; when the known points are point of the same order, the difference of point position or transverse displacement shall not be greater than 0.3mm on the map.

Table 4.4.1-1 Main technical requirements of traverse survey

Class	Length of traverse (m)	Mean square error of measured distance (mm)	The most number of traverse angles					Round number of horizontal angle observation			Mean square error of measured angles (″)	Azimuth closure (″)	Relative closure of total length of traverse	Length relative mean square error of the weakest side
			1:500 1:1000	1:2000	1:5000	1:10000		DJ1	DJ2	DJ6				
Third class	$15.0 \times M$	±15	15	15	25	35		6	9	—	±1.8	±$3.6\sqrt{n}$	1/60000	1/80000
Fourth class	$10.0 \times M$	±20	10	20	35	40		4	6	—	±2.5	±$5\sqrt{n}$	1/40000	1/40000
Fifth class	$4.0 \times M$	±30	10	20	35	40		—	3	6	±5.0	±$10\sqrt{n}$	1/20000	1/20000
	$2.5 \times M$	±30	6	15	23	25		—	2	4	±10.0	±$20\sqrt{n}$	1/10000	1/10000

Notes: 1. M is the denominator of mapping scale; n is the number of traverse angles.
2. When the mapping scale of surveyed area is larger than 1:1000, the length of traverse is determined according to the mapping scale of 1:1000.
3. The length of order 5 traverse in narrow difficult areas may be appropriately prolonged, but the number of traverse angles shall not be greater than the regulations of the table. The closure of class 5 traverse should not be greater than 0.3mm in the map.

Table 4.4.1-2 Main technical requirements of topographic control traverse survey

Number of development classes	Class	Length of traverse (m)	The most number of traverse angles		Azimuth closure (")		Mean square error of measured angles (")		Coordinate closure (mm) on the mappomg
			1:500 1:1000 1:2000	1:5000 1:10000	1:500 1:1000 1:2000	1:5000 1:10000	1:500 1:1000 1:2000	1:5000 1:10000	
Only one order	Frist order	$2.0 \times M$	15	30	$\pm 60\sqrt{n}$	$\pm 40\sqrt{n}$	± 30	± 20	0.40
Total two orders	Frist order	$1.5 \times M$	15	30					0.30
	Second order	$1.0 \times M$	10	20					0.26

Notes: 1. M is the denominator of mapping scale; n is the number of traverse angles.
2. The length of traverse in difficult areas may be appropriately prolonged. The allowed mean square error of point position of the last topograpic control point related to the adjacent basic horizontal control point is 0.1mm in the map.

4.4.3 The site selection for the control point of traverse network may be implemented in accordance with Article 4.3.2.

4.4.4 Besides the regulations of Article 4.3.3, the field work of traverse observation shall be in accordance with the following regulations:

1 Optical plummet of instrument shall be tested and rectified anytime. When horizontal angle of traverse is observed on the short side, instrument and sighting target shall be centered. Traverse of order three and order four should adopt three tripods method or forced centering base.

2 If, influenced by external factors, the compensator cannot work normally or works beyond the scope of compensation, observations shall be stopped.

4.4.5 Technical requirements of horizontal angle observation of traverse survey shall be carried out according to the Article 4.3.4.

4.4.6 If the survey station of the third and fourth class traverse has only two directions, the left angles shall be observed in the odd rounds and the right angles shall be observed in the even rounds. After the left and right angles adopt the median value, the misclosure is calculated with Equation (4.4.6), and then it shall not be greater than 2 times the mean square error of surveyed angle of corresponding order in the Table 4.4.1.

$$\Delta = [left\ angles]_{median} + [right\ angles]_{median} - 360°$$

(4.4.6)

4.4.7 The horizontal angle observation of topographic traverse may be observed with single round when DJ2 instrument is adopted, and with full round when DJ6 instrument is adopted, and the difference between clockwise and anti-clockwise round shall not be greater than 24″.

4.4.8 Technical requirements of range observation of traverse shall be implemented in accordance with Article 4. 3. 6 – 4. 3. 9.

4. 4. 9 Technical requirements of zenith angle observation of traverse shall be implemented in accordance with Article 4. 3. 10.

4. 4. 10 Correction and reduction of range observation of traverse shall be in accordance with regulations of Article 4. 3. 11.

4. 4. 11 Mean squared error of traverse angle observation may be calculated according to the angular misclosure of left and right angles or azimuth misclosure of traverse.

1 When the angular misclosure of left and right angles is used, the mean squared error of traverse angle observation is calculated according to Equation (4. 4. 11 – 1).

$$m_\beta = \pm \sqrt{\frac{[\Delta\Delta]}{2n}} \qquad (4.4.11-1)$$

Where:

m_β = mean squared error of traverse angle observation, ($''$);

Δ = misclosure of angles in a circular section of station, ($''$);

n = the number of misclosure of angles in a circular section of station.

2 When the azimuth misclosure of traverse is used, mean square error of traverse angle measurement may be calculated according to Equation (4. 4. 11 – 2).

$$m_\beta = \pm \sqrt{\frac{1}{N}\left[\frac{f_\beta f_\beta}{n}\right]} \qquad (4.4.11-2)$$

Where:

f_β = azimuth misclosure of annexed or closed traverse, ($''$);

n = the number of survey stations when f_β is calculated;

N = the number of annexed or closed traverses.

4. 4. 12 Accuracy evaluation of traverse distance measurement shall meet the following requirements:

1 Mean squared error of once distance measurement shall

be calculated with Equation (4.4.12-1).

$$m_0 = \pm\sqrt{\frac{[dd]}{2n}} \qquad (4.4.12-1)$$

Where:

$d =$ the difference between forward horizontal distance measurement and backward horizontal distance measurement of each side after reduced to the same height surface, mm;

$n =$ the number of bilateral observations.

2 Mean squared error of the average value of bilateral distance measurements shall be calculated with Equation (4.4.12-2).

$$m_d = \pm\frac{m_0}{\sqrt{2}} \qquad (4.4.12-2)$$

Where:

$m_d =$ Mean squared error of the distance measurement of once bilateral observations, mm.

4.4.13 The calculation of each order traverse and traverse network should use strict adjustment method. Accuracy evaluation shall include mean squared error of unit weight, and the parameters of point error ellipse, relative point error ellipse, etc.

4.5 Relevant Data Compilation and Submission

4.5.1 After completion of the horizontal control survey work, the following material should be compiled:

1 Technical design document.

2 Description of monument station.

3 Control network plot map.

4 Original record materials.

5 Horizontal control calculation data and control results table.

6 Test records/materials of survey instruments and tools.

    7 Technical summary reports.

    8 Other relevant material.

4.5.2 After completion of the project, the following documents should be submitted:

    1 Technical design document.

    2 Description of monument station.

    3 Control network plot map.

    4 Achievement table of control point.

    5 Technical summary reports.

5 Vertical Control Survey

5.1 General Provisions

5.1.1 The vertical control survey can be divided into basic vertical control, topographic vertical control, and station vertical control. The vertical control can be determined by means of leveling, trigonometric leveling with electromagnetic distance measurement (EDM) and GNSS Vertical survey. The vertical control order layout and accuracy requirements shall meet the requirements specified in Table 5.1.1.

Table 5.1.1 Technical requirements for arrangement, means and accuracy of vertical control network

Vertical control class		Vertical control order layout			Accuracy requirements
		$h=0.5m$	$h=1.0m$	$h\geqslant 2.0m$	
Basic vertical control	First	First, second class leveling			Permissible height RMSE of weakest point is $\pm h/20$. When $h=0.5m$, allowable RMSE is $\pm h/6$
	Second				
	Third	Third, fourth, fifth class leveling third, fourth, fifth class trigonometric leveling with EDM			
	Fourth				
	Fifth				
Topographic vertical control		The first order	The first order		Permissible height RMSE of height control points gotten in last densified height surveying to adjacent basic height control points is $\pm h/10$, and it shall not greater than $\pm 0.5m$
			The second order		
Station vertical control		The station	The station		Permissible height RMSE of station points to adjacent mapping height control points is $\pm h/6$
Note: h is basic contour interval of topographic map, m.					

5.1.2 The order of primary vertical control network shall be reasonably determined according to project scale, network purpose and accuracy requirements based on existing height control network. The primary height control network should be laid into an annexed leveling line, closed line or junction point network.

5.1.3 The vertical control line shall select existing higher order leveling point as known point.

5.1.4 If the vertical control point gotten by branch line leveling survey from national leveling point is taken as reference data in water and hydro-power station area vertical control, the accuracy shall comply with the following requirements: when the length of branch line leveling route is longer than 80km, the third class leveling survey accuracy requirements shall be adopted; when the length of branch line leveling route is less than 80km, the fourth class leveling survey accuracy requirements shall be adopted. Direct and reversed observation shall be adopted when branch line leveling survey is in process.

5.1.5 Significant digit in vertical control calculation result shall meet the requirements specified in Table 5.1.5.

Table 5.1.5 **Requirements of decimal digits in vertical control calculation result**

Class	Sum of direct and inverse distances (km)	Median of direct and inverse distances (km)	Edge of EDM (mm)	Zenith angle (″)	Elevation difference of each station (mm)	Sum of direct (inverse) distances (mm)	Elevation difference median of direct and inverse	Elevation
First	0.01	0.1	—	—	0.01	0.01	0.1	0.1
Second	0.01	0.1	—	—	0.01	0.01	0.1	0.1

Table 5.1.5 (Continued)

Class	Sum of direct and inverse distances (km)	Meidan of direct and inverse distances (km)	Edge of EDM (mm)	Zenith angle (")	Elevation difference of each station (mm)	Sum of direct (inverse) distances (mm)	Elevation difference meidan of direct and inverse	Elevation
Third	0.01	0.1	1	0.1	0.1	0.1	1	1
Fourth	0.01	0.1	1	0.1	0.1	0.1	1	1
Fifth	0.01	0.1	1	1	1	1	1	1
Mapping control	0.01	0.1	1	1	1	1	1	1

5.2 Site Selection and Monumentation

5.2.1 The vertical control route of each order shall be laid along roads or rivers with small slope, solid ground and convenient surveying condition.

5.2.2 The basic vertical control point shall be marked using permanent markstone. The permanent markstone also should be built on the bed rock or stable permanent building. In reservoir area, the permanent markstone shall be buried near residential area, mines, important cultural relics, road junction located near reservoir backwater pinch point or above normal storage water level line. The quantity of markstones of vertical control pass points shall be determined as appropriate. The existing markstones of the horizontal control points and fixed marks in a survey area may be employed as the markstones of vertical control pass points.

5.2.3 The markstone of vertical point shall be buried in the soil hard, stable, safe, secluded place where is convenient for long-term preservation and observation. The buried specifications for the markstone or mark of the vertical control point shall accord with the Annex B.

5.2.4 The description of station shall be drawn and the rough coordinates shall be collected on site after the markstone of vertical control point is buried.

5.2.5 The leveling route of fourth and above class should be set up a leveling markstone every 4 – 8km. The distance can be shorten to every 2 – 4km in hub project area or extended to 10km in desert area and branch leveling route. The distance between two buried markstones for the fifth class leveling may be determined as appropriate. But it should not be greater than 4km.

5.2.6 The leveling markstones shall spend a period of stability before vertical control survey field observation. The stability period of first and second order concrete leveling markstone is a rainy season or a frost free period. The stability period of first and second order rock leveling markstone is a month. After finishing the burial of markstone of third to fifth order, the starting time of observation shall be determined by measurement unit according to soil feature in the route and working season.

5.3 Leveling

5.3.1 The leveling route length of the first to fifth class shall not exceed the requirements specified in Table 5.3.1.

5.3.2 The main technical requirements of each class leveling shall not exceed the requirements specified in Table 5.3.2.

5.3.3 Topographic leveling route shall start and end on basic vertical control points. It shall be laid in the forms of annexed line, circuit or junction point network. Topographic leveling route may generate another topographic leveling route with the same accuracy. The line length and the differences of height differences of the two routes shall be less than the values as shown in Table 5.3.3.

Table 5.3.1 Requirements of leveling route length

Unit: km

Class	First	Second	Third		Fourth		Fifth	
Condition of survey area	—	—	$h \geqslant 1.0m$	$h = 0.5m$	$h \geqslant 1.0m$	$h = 0.5m$	$h \geqslant 1.0m$	$h = 0.5m$
Circuit perimeter	1500	750	200	50	100	20	45	16
Annexed line length	—	450	150	50	80	20	45	16
Branch line length	—	150	50	15	20	10	15	6
Junction point distance in the same order network	—	—	70	15	30	6	15	5

Notes: 1. When junction points are composed of routes, length of line between starting point and starting point shall not be greater than 1.5 times the values as shown in the table; length of line between starting point and junction point or between junction point and junction point shall not be greater than 0.7 times the values as shown in the table.
2. In difficult and mountain areas, leveling line length requirements of the first and second class may be relaxed according to concrete condition.
3. When skipping leveling orders and densified vertical control points, length of line requirements may be properly relaxed. But the elevation accuracy shall meet the requirements of Table 5.1.1.
4. h is basic contour interval, m.

Table 5.3.2 Main technical requirements for leveling

Unit: mm

Class	RMSE of height differences median per 1000m		Difference of height differences detection of measured segments	Difference of height differences between direct and inverse segments	Difference of height differences between left-right route	Misclosure of annexed or circuit leveling	Difference of height differences between direct and inverse in mountain area
	$M_\triangle$	M_W					
First	±0.45	±1.0	$3\sqrt{R}$	$1.8\sqrt{K}$	—	$2\sqrt{L}$	—
Second	±1.0	±2.0	$6\sqrt{R}$	$4\sqrt{K}$	—	$4\sqrt{L}$	—
Third	±3.0	±6.0	$20\sqrt{R}$	$12\sqrt{K}$	$8\sqrt{K}$	$12\sqrt{L}$	$4\sqrt{n}$ or $5\sqrt{L}$
Fourth	±5.0	±10.0	$30\sqrt{R}$	$20\sqrt{K}$	$14\sqrt{K}$	$20\sqrt{L}$	$6\sqrt{n}$ or $25\sqrt{L}$
Fifth	±7.5	±15.0	$40\sqrt{R}$	$30\sqrt{K}$	$20\sqrt{K}$	$30\sqrt{L}$	$10\sqrt{n}$ or $40\sqrt{L}$

Notes: 1. $M_\triangle$, M_W respectively represent accident RMSE of height differences median per 1000m and total RMSE of height differences median per 1000m, mm.
2. R is length of detection section, km; K is length of route, section or segment, km; L is length of annexed or circuit leveling line, km; n is the number of stations. When $R<1$km, the value of R is 1km; When $L<100$m, the value of L is 100m.
3. The tolerance of "difference of height differences detection of measured segments" is suitable for single leveling line detection and duplicate leveling line detection.
4. When one-way leveling station number $n>16$, the calculation of differences of height differences may accord to n.
5. When leveling circuit route is composed of different order leveling routes, the tolerances of circuit leveling misclosure shall be calculated respectively according to different order leveling line lengths, and the tolerance is the square root of sum of squares of them.

Table 5.3.3 Requirements of line length and accuracy of topographic control leveling

Basic contour interval (m)	Total RMSE of height differences median per 1000m (mm)	Length of annexed or circuit leveling line		Misclosure of elevation (mm)	
		Primary mapping control leveling	Second mapping control leveling	Even ground, undulating ground	Mountainous ground
0.5	±20	12	12	±40 $\sqrt{L}$	±12 $\sqrt{n}$
1.0 and above	±20	30	30	±40 $\sqrt{L}$	±12 $\sqrt{n}$

Notes: 1. L is the length of annexed or circuit leveling line, km; n is the number of stations.
2. When junction points are composed of routes, length of line between starting point and starting point shall not be greater than 1.5 times the values as shown in the table; length of line between starting point and junction point or between junction point and junction point shall not be greater than 0.7 times the values as shown in the table.

5.3.4 The instruments and calibration requirements, observation methods, observation tolerances, data processing method of the first, second third and fourth class leveling shall comply with GB/T 12897, GB/T 12898.

5.3.5 The observation sequence of the fifth class leveling shall be Back-Back-Fore-Fore. The distance can be observed directly by reading by optical level cross-hair observation method. The fifth order annexed line or circuit route may adopt one-way observation. The fifth branch leveling may adopt direct and reversed observation method or single-pass double turning point observation method.

5.3.6 The length of sight, difference of fore and back sight distance, height of sight from ground of the fifth class leveling meet the requirements specified in Table 5.3.6.

Table 5.3.6 Requirements of sight distance, difference of fore and back sight distance, height of sight for the fifth class leveling

Type of level	Maximum sight distance (m)	Difference of fore and back sight distance (m)	Accumulated difference of fore and back sight distance (m)	Height of sight from ground	Number of repeated measurement
Optical level (DS3)	150	≤20.0	≤100.0	Three wire can be read	—
Digital level (DSZ3)	100	≤20.0	≤100.0	Can be read	≥2

5.3.7 The each station observation tolerances of the fifth order leveling shall not exceed the requirements specified in Table 5.3.7.

Table 5.3.7 Station tolerances requirements for the fifth order leveling   Unit: mm

Order	Observation method	Differences of red and black side readings or twice readings	Difference of height differences of red and black side readings or twice readings	Turning points reading differences of left-right route	Difference of height differences of detection interval points
Fifth	Cross-hair observation	4	6	6	6

5.3.8 When the fifth class leveling route crosses rivers by using general observation method, maximum sight distance shall not be greater than 250m. The observation shall be done twice with instrument height changed and the height differences of twice observation shall not be greater than 14mm.

5.3.9 When crossing river sight distance in the fifth class leveling route exceeds the maximum sight distance of general observation method, the leveling method, the distance, observation set number, and tolerances shall meet the requirements specified in Table 5.3.9.

Table 5.3.9 Station tolerances requirements for the fifth class cross-river leveling

Method	Maximum sight distance S (m)	Single observation set number	Observation group number of half observation set	Difference of height differences of observation sets (mm)
Directly reading method	0.4	2	—	$\leqslant 24$
Jiggling target method	1	4	—	$\leqslant 60S$
Tilting method with transit or trigonometric leveling with EDM	2	8	3	$\leqslant 6\sqrt{S}$

5.3.10 When the fifth class leveling route crosses river with flat water, hydrostatic lake, pond stagnant water area and reservoir with the width exceeding 300m, quiescent water level transfer height survey method may be adopted and shall comply with the following requirements:

1 When leveling route crosses river, the route shall cross the stream segment of tributary and shall not cross multi flow channel (multi water flow) stream section.

2 When leveling route crosses quiescent water area, the route shall be selected in a small span location with small external disturbance.

3 The height transfer survey using quiescent water level shall be done at calm and breezeless moment, the height differences between water level of both sides and the pile's top level shall be measured at the same time.

4 The whole work shall be carried out twice. The tolerance of the difference of two measurement results is 30 $\sqrt{L}$, and L is the distance between two mark stones, km.

5.3.11 Topographic leveling shall use the level tools whose precision is no less than that of DS10 level and two-sided leveling rod with circular bubble. Topographic leveling shall use cross-hair observation method in single leveling line and the observation data shall be estimated to mm. The tolerances of station observation shall meet the requirements specified in Table 5.3.11.

Table 5.3.11 **Main technical requirements for topographic leveling**

i angle (″)	Length of sight (m)			Differences of fore-back sight distance	Accumulated difference of fore-back sight distance	Differences between red and black side readings differences and constants	Differences between height differences of red side and black side
	General imaging quality	Clear imaging quality	Crossing river				
30	100	150	200	20	100	4	6

5.3.12 Each completed leveling route surveyed by direct and reversed observation or left-right route shall calculate discrepancy of height differences and accident RMSE per 1000m M_Δ except for the fourth and fifth class single leveling route and mapping leveling route. The calculation shall meet the requirements specified in Table 5.3.2. The route, in which the length is less than 100km or the segments is less than 20, may be incorporated in adjacent route and calculated together. Accident RMSE per 1000m M_Δ shall be calculated according to Equation (5.3.12):

$$M_\Delta = \pm \sqrt{\frac{1}{4n}\left[\frac{\Delta\Delta}{R}\right]} \qquad (5.3.12)$$

Where:

Δ = discrepancy between the height differences of direct and reverse segments (or left-right route) of leveling, mm;

R = spacing distance of leveling segment, km;

n = number of segments.

5.3.13 Each completed annexed line or circuit leveling route shall correct observation height differences result including leveling rod length error and nonparallel normal horizontal plane correction and then calculate misclosure of annexed line or circuit route. The calculation shall meet the requirements specified in Article 5.3.2 and Article 5.3.3.

5.3.14 When the number of closed leveling routes in a leveling network exceed 20, the total RMSE of mean height differences per 1000m M_W shall be calculated according to the misclosure of closed leveling route W and meet the requirements specified in Table 5.3.2 and Table 5.3.3. Total RMSE of mean height differences per 1000m M_W shall be calculated according to Equation (5.3.14):

$$M_W = \pm \sqrt{\frac{1}{N}\frac{WW}{F}} \qquad (5.3.14)$$

Where:

W = the misclosure of closed leveling route after correction, mm;

F = perimeter of closed leveling route, km;

N = number of closed leveling routes.

5.3.15 Leveling network shall be processed with rigorous adjustment and evaluation of precision.

5.4 Trigonometric Leveling with EDM

5.4.1 The third, fourth, fifth class and topographic trigonometric leveling with EDM should be laid out and measured together with horizontal control survey, and may also be laid out in the forms of an annexed height traverse (closed height traverse) or height traverse network.

5.4.2 Each order trigonometric leveling with EDM route shall start and end on higher order control points from leveling or trigonometric leveling with EDM.

5.4.3 Every side and whole line length of each order of trigonometric leveling with EDM shall not exceed the requirements specified in Table 5.4.3.

Table 5.4.3 Side and route length requirements for the third, fourth and fifth class trigonometric leveling with EDM

Class	Length of each side (m)	Route length (km)	
	Reciprocal observation	$h=0.5$m	$h \geqslant 1$m
Third	700	40	120
Fourth	1500	20	60
Fifth	2000	15	40
Mapping	1300	7	30

Notes: 1. h is basic contour interval, m.
2. When junction points are composed of routes, length of route shall comply with "Note 1" in Table 5.3.1.

5.4.4 Operation methods of each order trigonometric leveling with EDM shall meet the requirements specified in Table 5.4.4. Observation method of meteorological data shall comply with Article 4.3.6.

Table 5.4.4 Operation methods of trigonometric leveling with EDM

Order	Measure method	Route	Meteorological data		Level of EDM	Side Observation set number		Level of angular instrument	Zenith angle Observation set number	
			Observation time interval	Data acquisition		Direct	Inverse		Cross-hair observation	Three wire observation
Third	Reciprocal observation	Double observation of single line	Beginning and end of each side observation	Average of two endpoints of each side	I II	2 4	2 4	DJ1 DJ2	2 4	1 2
Fourth	Reciprocal observation	Single line	Beginning and end of each side observation	Average of two endpoints of each side	II III	2 4	2 4	DJ1 DJ2	2 4	1 2
Fifth	Reciprocal observation	Single line	1 times each side	Data on station	II III IV	2 2 4	—	DJ2	2	1
Mapping	Reciprocal observation	Single line	1 times each side	Data on station	II III IV	2 2 4	—	DJ2 DJ6	2 4	1 2

Notes: 1. An observation set of side measurement is read 4 times at one aiming point.
2. Double observation of single line is 2 times observation with changing height of instrument or height of target or prism.

5.4.5 Zenith angle shall be observed using cross-hair or three wire observation methods by total station, electronic theodolite or theodolite. Aiming at the same target and reading by face left and face right respectively is called observation set. When horizontal hair is used to aim at target, the target shall be accurately aimed at 2 times and 2 readings shall be gotten every time. The tolerance of difference of 2 readings is 3″.

5.4.6 Slope distance shall be measured by total station or range finder. When total station is used in observation, instrument parameters shall be pre-set. Observation steps of the same target in an observation set at a station shall comply with the following requirements:

1 Instrument, prisms, barometer, thermometer shall be placed in dry and shady place for 15min at least. The thermometer shall be hung in the shade with the same height of the instrument. The thermometer shall be put on a horizontal plane and the pointer shall not be hindered.

2 Level the instrument and prism. Measure and record the height of instrument and prism.

3 Aim at prism and measure slope distance for 4 times. Record values of slope distance, temperature and pressure.

5.4.7 Observation tolerances of each order of trigonometric leveling with EDM shall meet the requirements specified in Table 5.4.7.

5.4.8 Each order of trigonometric leveling with EDM shall also comply with the following requirements:

1 Distance measurement with EDM shall comply with requirements of side measurement in Section 4.3.

2 Reciprocal observation shall be finished in a short time.

3 Zenith angle should be measured simultaneously with re-

Table 5.4.7 Observation tolerances of trigonometric leveling with EDM

Order	Side measurement (mm)			Zenith angle measurement (″)		Height difference				
	Difference of readings in an observation set	Difference between observation sets	Difference of direct and inverse	Difference of index errors	Difference between observation sets	RMSE of height differences median per 1000m	Difference of reciprocal height differences	Difference of single line duplicate observation height	Misclosure of annexed traverse (closed traverse)	Difference of height differences detection of measured segments
Third	5	7	2 $(a+bD)$	5	5	±6	35D	±8$\sqrt{[D]}$	±12$\sqrt{[D]}$	±20$\sqrt{[D]}$
Fourth	10	15	2 $(a+bD)$	8	8	±10	45D	±14$\sqrt{[D]}$	±20$\sqrt{[D]}$	±30$\sqrt{[D]}$
Fifth	10	15	—	10	10	±15	60D	±20$\sqrt{[D]}$	±30$\sqrt{[D]}$	±40$\sqrt{[D]}$
Mapping control	20	30	—	15	15	±20	75D	±25$\sqrt{[D]}$	±40$\sqrt{[D]}$	±50$\sqrt{[D]}$

Notes: 1. The slope distances shall be turned into the same plane before calculating the difference of direct and inverse.

2. $(a+dD)$ is the nominal accuracy of EDM.

3. D is the distance between two stations, km.

ciprocal observation in desert, water network areas and torridity, dramatic temperature changes areas.

    4 Zenith angle of each side in a route should not be less than 75° or greater than 105°.

    5 The trigeminy tripod method should be used in operation.

5.4.9 When some discrepancies of the height differences of reciprocal observations have certain regularity in a survey area and most of them approach or exceed the limit, the K value shall be calculated according to Equation (5.4.11 - 1) and Equation (5.4.11 - 2). The differences of direct and inverse observations shall also be recalculated.

5.4.10 Recalculating and accepting or rejecting the results of trigonometric leveling with EDM shall comply with the following requirements:

    1 All the results exceeding the requirements specified in 5.4.7 shall be remeasured.

    2 When zenith angle is measured using three wires method, if the zenith angle or difference of index errors measured by a horizontal hair in a certain direction exceed tolerances, it shall be remeasured by cross hair method in an observation set in this direction. If the results measured by two horizontal hairs in an observation set in the same direction exceed tolerances, the results shall be remeasured by three wires method in an observation set in this direction or be remeasured by twice observation sets using three-wires method.

    3 Recalculating and accepting or rejecting the results of trigonometric leveling with EDM shall comply with the requirements specified in Article 4.3.8.

5.4.11 The slope distance of the third, fourth, and fifth class of trigonometric leveling with EDM shall be first corrected by meteorological data, additive constant, and multiplication constant. Then, the following items may be calculated:

1 The K value of vertical refraction coefficient of atmosphere in a survey area shall be calculated according to Equation (5.4.11-1) or Equation (5.4.11-2) as appropriate.

1) Using reciprocal height difference:

$$K = 1 + \frac{R}{D_{AB}^2}[(S_{AB}\cos Z_{AB} + S_{AB}\cos Z_{BA}) + (i_A - l_B) + (i_B - l_A)]$$

(5.4.11-1)

Where:

K = atmospheric vertical refraction coefficient;

S_{AB} = corrected slop distance between station A and station B, m;

D_{AB} = horizontal distance between station A and station B, m;

Z_{AB} = zenith angle between station A and station B, (°);

i_A, i_B = instrument heights of station A and station B, m;

l_A, l_B = target heights of station A and station B, m;

R = mean radius of curvature of the Earth, m.

2) Using known height difference:

$$K = 1 + \frac{R}{D_{AB}^2}[(S_{AB}\cos Z_{AB} + i_A - l_B) - h_0]$$

(5.4.11-2)

Where:

h_0 = known height difference between station A and station B, m.

2 Height difference of single-way observation shall be calculated according to Equation (5.4.11-3).

$$h = D_{AB}\cot Z_{AB} + i_A - l_B + \frac{1-K}{2R}D_{AB}^2$$

(5.4.11-3)

3 Height difference of reciprocal observations shall be calculated according to Equation (5.4.11-4).

$$h = \frac{1}{2}\left[D_{AB}\cot Z_{AB} - D_{AB}\cot Z_{BA} + (i_A - l_B) - (i_B - l_A) - \frac{K_{AB} - K_{BA}}{2R}D_{AB}^2\right] \quad (5.4.11-4)$$

Where:

K_{AB} = atmospheric vertical refraction coefficient from station A to station B;

K_{BA} = vertical refraction coefficient of atmosphere from station B to station A.

4 Random RMSE of height differences per 1000m M_Δ shall be calculated with discrepancies of double measured mean height differences of each side according to Equation (5.4.11-5).

$$\left.\begin{array}{r}M_\Delta = \sqrt{\frac{[Pdd]}{4n}} \\ P = \dfrac{1}{S^2}\end{array}\right\} \quad (5.4.11-5)$$

Where:

d = discrepancy of height difference, mm;

n = number of discrepancies;

S = slope distance, km.

5 Total RMSE of mean height differences per 1000m M_W shall be calculated with discrepancy height difference of each closing circuit according to Equation (5.3.14).

5.4.12 Trigonometric leveling with EDM for vertical topographic control may be developed continuously twice. The length of each route shall not exceed the requirements specified in Table 5.4.3.

5.4.13 Slope distance, horizontal distance and height difference of trigonometric leveling with EDM may be measured directly. The stipulations specified in Table 5.4.13.

Table 5.4.13 Technical requirements for mapping and station trigonometric leveling with EDM by total station

Order	Meteorological elements	Instrument nominal accuracy		Observation sets of slope distance and height difference	
		Ranging accuracy (mm/km)	Accuracy of angular measurement (″)	Face left	Face right
Mapping	Input measured elements to the instrument before observation; if temperature change exceed 1℃, elements shall be reentered for each observation set	±5	2	2	2
		±10	7	4	4
Station	Input measured elements to the instrument before observation	±5	2	1	1
		±10	7	2	2

5.4.14 Height differences of benchmarks or other fixed points measured by the third, fourth and fifth class trigonometric leveling with rangefinder shall be corrected to consider the unparallel impact of geoidal surface, and the calculation method is the same as leveling.

5.4.15 When the third, fourth, fifth class and mapping trigonometric leveling with EDM routes forms a height network, it should be processed using rigorous adjustment and evaluation of precision.

5.5 GNSS Vertical Survey

5.5.1 GNSS vertical survey may adopt GNSS fitting leveling, RTK height survey, GNSS leveling based on refined geoid model, GNSS cross-river leveling and etc. The application scope of various methods mentioned above shall meet the requirements specified in Table 5.5.1.

Table 5.5.1 Application scope of GNSS vertical survey methods

GNSS leveling methods	First, second, third order	Fourth order	Fifth order	Mapping	Station
GNSS fitting leveling	—	—	Yes	Yes	Yes
RTK height survey	—	—	—	Yes	Yes
GNSS leveling based on refined geoid model	—	Yes	Yes	Yes	Yes
GNSS cross-river leveling	Yes	Yes	Yes	Yes	Yes

5.5.2 Main technical requirements for GNSS fitting leveling shall comply with the following requirements:

1 GNSS fitting leveling should be done with GNSS horizontal control survey at the same time or can be done separately.

2 Connecting survey of GNSS network shall be done with fourth or higher class leveling points. GNSS connecting survey points should be distributed around and in the center area of a survey area. The number of GNSS connecting survey points should be greater than 1.5 times the number of unknown parameters in selected fitting model. The point spacing should be less than 10km.

3 The number of GNSS connecting survey points shall be increased according to terrain features of a survey area with large height differences.

4 Partition fitting method should be adopted in a large survey area with obviously changed terrain trend. 2 - 3 coincident points should be selected in each of the partitions.

5 GNSS fitting leveling shall adopt dual-frequency receiver with fixed error no more than 10mm and ratio error coefficient no more than 2mm/km. Observation technique shall comply with the requirements of corresponding order in Section 4.2. The number of sessions shall not be less than 1.6 and the session observations shall be appropriately extended according to the actual situation. The antenna height shall be measured twice before and after observation. The mean of two antenna heights shall be taken as final result if the difference is less than 3mm.

5.5.3 GNSS fitting leveling calculation shall comply with the following requirements:

1 Make full use of local gravimetric geoid model and data.

2 Reliability test of known connecting survey points shall be done and unqualified point shall be removed.

3 Plane fitting model shall be adopted in a small area with flat terrain. Surface fitting model shall be adopted in a large area with large topographical undulation.

4 Height fitting model shall be optimized.

5 Fitting height points should not exceed the range of area covered by known points.

6 Height fitting model selected shall calculate model interior coincidence RMSE. Model interior allowable coincidence RMSE of fifth order shall be ± 2cm and model interior allowable coincidence RMSE of mapping and station order shall be ± 3cm.

Model interior allowable coincidence RMSE M_h shall be calculated according to Equation (5.5.3-1):

$$M_h = \pm \sqrt{\frac{[d_h d_h]}{n-1}} \qquad (5.5.3-1)$$

Where:

M_h = height anomaly model interior coincidence RMSE, cm;

d_h = difference between fitting point leveling elevation and model calculation elevation, cm;

n = number of fitting points.

7 Height fitting model selected shall calculate model exterior coincidence RMSE. Model exterior allowable coincidence RMSE of fifth order shall be ±3cm and model exterior allowable coincidence RMSE of mapping and station order shall be ±5cm. Model exterior allowable coincidence RMSE M_h shall be calculated according to Equation (5.5.3-2):

$$M_h = \pm \sqrt{\frac{[d_h d_h]}{n}} \qquad (5.5.3-2)$$

Where:

M_h = exterior coincidence leveling RMSE, cm;

d_h = difference between the elevations checked with leveling and calculated with model, cm;

n = number of checking points.

5.5.4 GNSS height fitting result shall be checked. Number of checking points shall not less than 10% of all the fitting points and shall not be less than 3. Height difference check may adopt corresponding order leveling method or trigonometric leveling with EDM method. Permissible height difference shall be $40\sqrt{D}$ mm in the fifth order and $50\sqrt{D}$ mm in topographic order and station order. D is length of check route, and is measured in km.

5.5.5 Vertical survey using GNSS RTK shall comply with the following requirements:

1 RTK height control points should be buried synchronously with GNSS RTK horizontal control points. The markstones of them can be the same. If the same markstones are used for horizontal and Vertical control points, the marks shall be bulletheaded and with crossing hair.

2 Main technical requirements for Vertical control survey using RTK shall meet the requirements specified in Table 5.5.5.

Table 5.5.5 **Main technical requirements for Vertical control survey using RTK**

Geodetic height RMSE (cm)	Distance to the base station (km)	Observation number	Reference point order
$\leqslant \pm 3$	$\leqslant 5$	$\geqslant 2$	The fifth order and above leveling point

Notes: 1. Geodetic height RMSE is the error of control geodetic height relative to the nearest base station.
2. Network RTK height control survey may not limited by the distance between moving station to base station, but shall be in the effective range of services.

3 Setup of rover station and base station shall comply with Article 4.2.10.

4 Each time before RTK operation or after base station re-setup, a check shall be done by at least one known point at the same order or higher order known point. Allowable height difference shall be 10cm.

5 Allowable convergence error of the height control point determined by RTK survey shall be 3cm.

6 Allowable height difference of each survey shall be 1/10 of basic height interval and the median value shall be taken as the final result.

7 The determination of height control point with RTK may be gotten by geodetic height measured by rover station minus the height anomaly of the rover station.

8 Anomaly of the rover station height may be gotten by mathematical fitting method and refined quasi geoid. Accuracy of model fitting quasi geoid shall be determined according to actual production needs. Mapping RTK moving station height anomaly may be gotten by point adjustment method in scene of survey area. Allowable residual error of mapping control point height fitting shall be 1/10 of basic height interval.

5.5.6 The relative height difference between the control points shall be calculated in GNSS height control survey based on refined geoid model. The calculation shall be started from higher order control point and accorded to annexed leveling route. The misclosure of the route shall meet the corresponding order requirements.

5.5.7 GNSS cross-river leveling shall comply with the related requirements specified in GB/T 12897 and GB/T 12898.

5.6 Relevant Data Compilation and Submission

5.6.1 After completion of vertical control survey, the following materials shall be compiled:

  1 Technical design document.
  2 Description of survey monument.
  3 Height control network diagram.
  4 Original data.
  5 Height control calculation data and control points result table.
  6 Inspection documents of various survey instruments and

tools.
    7 Technical summary report.
    8 Other relevant materials.

5.6.2 After the completion of the project, the following materials shall be submitted:
    1 Technical design document.
    2 Description of buried markstones.
    3 Vertical control network diagram.
    4 Control points result table.
    5 Technical summary report.

6 Digital Topographic Map Surveying

6.1 General Provisions

6.1.1 Digital topographic map surveying and mapping can use total station and RTK surveying method. Field data acquisition may use coding method, sketch method and integration method etc.

6.1.2 The density of terrain points to reflect the terrain changes for the principle, the largest spacing of topographic point should not be greater than the provisions of the Table 6.1.2.

Table 6.1.2 **Topographic point maximum spacing**

Scale	1:500	1:1000	1:2000	1:5000
Point spacing (m)	15	30	50	100

6.1.3 Digital topographic map surveying and mapping should be controlled, following the principle of do not paint if it cannot be seen clearly.

6.2 Digital Mapping

6.2.1 The following preparation work should be undertaken before digital map:

    1 Set up to distinguish the image information, such as sheet number, border point coordinate range, etc.

    2 Calibration for results of control points.

6.2.2 Digital mapping instrument settings and station inspection shall meet the following requirements.

    1 Total station mapping instrument placement and station check shall meet the following requirements:

        1) Instrument for the centering error is not more than 5

mm. The instrument height and prism height should amount to 1cm.

2) The fruther mapping control point shall be served as the directional point, and the coordinates and elevation of other mapping control point as check point shall be checked. The horizontal position permissible discrepancy of the check point on the map shall be 0.2mm, and the vertical permissible discrepancy shall be 1/5 of basic contour interval.

3) The directional azimuth shall be checked during the course of the operation and before the end.

2 If the single base station is used for surveying, the distance between the RTK surveying rover and the base station shall not be further than 5 km. If the Network RTK is used for surveying, the distance between the RTK surveying rover and the base station may not be restricted, but the surveying area shall be in the effective service scope with network coverage. Before the implementation of survey, not less than two known points which hold higher precision than mapping control point shall be checked, and the tolerance of detection result shall be equal to the total station mapping.

6.2.3 Station measurement shall meet the following requirements:

1 When density of control point in mapping is not enough, station point may be added using open traverse method, polar coodinate method, freely set station method. The point position RMSE of station point relative to adjacent mapping base point shall be 0.2mm in the map, elevation allowable error shall be 1/6 of the contour interval.

2 When the total station has been used to measure the coordinates and elevation of station point directly, the average of the observed values before semi-observation and after semi-observation shall be calculated as the final observation results, and side length should not be greater than the maximum length of ranging topographic point in Table 6.2.4.

3 When the measurement was evaluated by method of free site coordinates and elevation, the number of known points in the observation should not be less than 2. When both groups were observed, coordinates and elevation allowable difference shall not exceed $2\sqrt{2}$ times the station measurement accuracy requirements in Table 4.1.2 and Table 5.1.1.

4 When RTK has been used for laying out station points, the observation shall be done two times, and the horizontal permissible discrepancy shall be 0.5mm in the map, and the vertical permissible discrepancy shall be 1/3 of the basic contour interval.

6.2.4 When using the total station mapping, operations shall be done according to the following requirements:

1 Total station mapping range finder length should not exceed the provisions of the Table 6.2.4.

Table 6.2.4 **The maximum length of distance measuring total station mapping rules**

Scale	Maximum range length (m)	
	Feature points	Topographic point
1:500	160	300
1:1000	300	500
1:2000	450	700
1:5000	700	1000
Note: For 1:2000 and 1:5000 scale maps, when basic contour interval is 0.5m, the maximum range should not be greater than 500m length.		

2 Total station mapping shall meet the following requirements:

1) When using the sketch method, the sketch should be drawn in accordance with the station, and the number of measuring points one by one. Number of measuring points should be consistent with the instrument records the dot. The sketch map, appropriate simplification of terrain factors of location, attributes, and relationships, etc.

2) When using coding method, it is appropriate to use general coding format as well as software custom function and extend the functionality to establish user's coding system.

3) When integration means mapping is used, it is necessary to establish real—time measuring point of the attribute, connection and logical relations, etc.

4) When the instrument without antomatic recording function has been used, horizontal angle and vertical angle observation should be read to seconds, distance should be read to centimeters, and coordinates and elevation calculation shall be accurate to cm.

5) When the measurement has been implemented according to mapsheet, each image should be measured outside the border line 5mm. When the measurement has been implemented according to partition measuring, it should be measured on the outer boundary of the area 5mm.

6) The collected data should be checked, processed, deposited in the disk to save after making sure that there is no mistake data, and all date is saved and

backed up.

3 Sketch map compilation shall meet the following requirements:

1) Sketch symbolization should be carried out according to the existing national topographic map schematism. It could be simplified and self definition for the cartographic symbolization. The coding used for data collection shall be one-to-one correspondence with the symbols in the sketch mapping.
2) The number of measuring points should be consistent with the data collection records.
3) Sketch of the positional relationship between the elements of terrain should be right and clear.
4) Various names and feature attributes in topographic maps should be noted clearly in the sketch.

6.2.5 When RTK surveying is carried out, in addition to meet the requirements of Article 6.2.4, it should also meet the following requirements:

1 Reference station of the effective radius of operation shall not exceed 5km.

2 Known points shall be checked before and after work, and the horizontal position discrepancy of check point shall not be greater than 0.2mm in the map. While the vertical position discrepancy shall not be greater than 1/5 of the basic contour interval.

3 In the process of operation, it should be initialized again if the satellite signal loses lock. It can continue to work after coincidence point measurement is found to be qualified by inspection.

4 While different moving stations, working in different

partitions of the border of measurement range shall be 5mm beyond outside line in the map.

5 When different reference station during operations has been used, a certain number of feature points shall be checked, and horizontal and vertical discrepancy of points shall not be greater than $2\sqrt{2}$ times the value provided in clauses 4 and 5 in Article 3.0.5.

6 After daily measurement, the collected data shall be deposited in computer, and back-up of data shall be done.

6.2.6 Digital terrain measurement features of surveying and mapping shall meet the following requirements:

1 Features of surveying and mapping shall include the following contents:

1) Control points.
2) Residential areas.
3) Water system and its annex.
4) Roads and pipelines.
5) Transmission lines and Communication lines.
6) Independent ground features.
7) Geological exploration points and hydrologic meteorological facilities.
8) The border line, the boundary of the class and fence.

2 When measured in accordance with the scale features, main contour points are measured with instrument, and the rest of the measurement is conducted by hand. When measuring half proportional or independent symbol mapping feature, it should measure its center position or anchor point.

3 When the measurement zone features is complicated, a proper choice may be done, the feature and orientation related to water and hydropower projects shall be retained particularly, and the geographical characteristics of the region shall be reflected. If the symbol is so intensive that all local things could not be accommodated totally, the main symbols should not change its position and secondary symbol may be moved or reduced appreciably, while a corresponding position characteristics shall be maintained. What's more, individual secondary symbols may be omitted.

4 Residents, important public buildings, rivers, lakes, mountains and other important features of the names should be investigated and noted. The note should be chosen in the appropriate location, and should not cover important features or landscape.

5 When mapping the residents, the outline of building is based on the wall.

6 Transmission lines, communication lines and one section of the pipeline that is set up on the ground or its part is buried, shall be surveyed and mapped in the map. The lines and underground pipelines whose directions could be determined shall be expressed with dotted line on the chart. The temporary lines and pipes may not be surveyed and mapped.

7 For the geologic exploration points such as drilling, exploratory trench, wells, and horizontal hole, in addition to the surveying and mapping as stipulated in the technical specification, it is feasible to plot them according to the existing coordinate and note their elevation. It can choose to map when geological exploration points are too close.

8 When the water system is being surveyed and mapped, if the water depth is greater than 1m or width of water level is greater than 5m, a point of water level shall be marked every 10 -15cm on the map, and the date of survey shall be indicated. The point of water level shall be surveyed and marked in the great variation of water level such as upstream and downstream of the beach, barrage, rivers confluence, towns and settlement place, bridges, ferry and other feature points. The point of water level shall be surveyed and marked near the mapborder, and the elevation of river bottom and channel bottom should be surveyed and marked for creek, stream and channel.

6.2.7 Geomorphology, soil and vegetation mapping shall meet the following requirements:

    1 Contour matching geomorphic signs and elevation note describes topography.

    2 When drawing contours, 1/2 contour should be added if basic contour cannot describe the saddle, hill, steps and the basin and other fine features. If the basic contour and half contour are still unable to express certain necessary topographical features, 1/4 contour shall be added.

    3 On the top of the hill, saddle, depression or terrain gradient where direction is not easy to distinguish, line slope should be drawn.

    4 When the interval of less than 2.5 mm on the drawing, the first curve may not draw.

    5 Steep mountainsides can be represented by steep cliffs or rock symbol. When the contours can be expressed, it is appropriate to represent by the first curve or curve, or a combination of the two.

    6 In addition to using contour rendering terrain ups and downs, it should also note elevation at the top of the mountain, saddle, terrace, concave, outstanding earth hummock and independent hillock, the ground slope change, karst cave mouth, fountain of outcrop, etc.

    7 The symbol shall be used to describe the landform elements, such as the steep cliff, independent stone, gully, rain crack, mound and terraces, potholes, embankment, terraced fields, etc. The elements should be noted if the elevation difference is greater than the basic contour interval.

8 For outcrop, independent stone, terraced fields should be measured and recorded. If the relative eleration difference of the slope or scarp is less than half the basic contour interval or the length is less than 5mm on the map, the elements may be not expressed on the map. If the slope or scrap is dense, the elements may be expressed or abandoned appropriately.

9 The following soils and vegetation elements should be plotted on the topographic map:
    1) Fields, orchards and nurseries.
    2) Grassland, wasteland and reeds.
    3) Forest, shrubs, and bamboos.
    4) Swamps and saline-alkali land.
    5) Rock, gravel, sand and gravel.

10 If the area of economic soil and vegetation is larger than 1 cm^2 on the map, the classification land boundary shall be drawn. The corresponding symbol for the fixed crop, cash crops and aquatic plants shall be used. The cultivated land for crop rotation should be distinguished by paddy and dry land. If a variety of plants is planted in the same plot, the design symbols including the soil symbol shall not exceed 3 kinds.

6.2.8 Topographic map editing and processing shall meet the following requirements:

1 Data processing of the original observation shall be implemented, including data classification, compilation and modification of data property, but the measured data shall not be modified.

2 Topographic map elements should be layered representation. Layered method and the layer named shall be implemented according to the provisions of Article 3.0.7. According to the project need the layer structure may be modified. But the entity of same layer should have the same color combination and attribute structure.

3 A variety of ground line, broken line and micro – landform representation should be put forward according to the authenticity of the geomorphic requirements when the digital terrain model is established.

4 The drawing and editing of various features, landscape symbols and notes shall be implmented according to the requirement of the Articles 6.2.6 and 6.2.7. When different properties of line overlap, it can be drawn together, and layers of different color shall be adopted.

5 When the topographic map subdivision is implemented, in addition to meet the Article 3.0.5, the following requirements shall be met at the same time.

1) The topographic map surveyed with several regions, cutting of mapsheet shall be implemented, and the margin data shall be inspected and compiled.

2) The topographic map surveyed in the type of mapsheet shall be implemented with map interlinking inspection and margin data compilation inspection. The permissible tolerance of edge matching shall be $2\sqrt{2}$ times of the values specified in Clauses 4 and 5 of Article 3.0.5. The average distrilution shall be implemented if the error conform to the tolerance, or the site inspection and compilation shall be done.

6 The compilation and inspection of topographic map shall

include the following contents.
1) Whether the connected relation of graphics is correct, whether the graphics is consistent with the sketch, and whether the graphics have error and omission.
2) Whether the location of the various annotation is appropriate, whether the annotations have kept away from the objects, symbols.
3) Whether the connection, intersection or overlap of various segments is appropriate and accurate.
4) Whether the drawing of the contours is coordinated with the ground line, whether the annotation is appropriate and the disconnect part is reasonable.
5) Whether the treatment is appropriate for the spacing which is less than 0.2 mm in the map with different attributes of the line.
6) Whether the relevant attribute information assignment of terrain and landform is correct.

7 The appropriate scale map of the terrain shall be printed after the completion of topographic map compilation process, and the inspection indoor and site inspection roundly shall be implemented according to the sample map. The problems shall be disposed in time if the problem is found.

6.3 Digital Topographic Map

6.3.1 Professional equipments used for digital topographic map shall meet the following requirements:

1 The resolution of the scanner should not be less than 12point /mm (300 dpi); Effective scanning area should not be less than 841mm×597mm (A1).

2 Graph plotter resolution that is used to check output

should not be less than 10 point/mm (254 dpi). Effective drawing area should not be less than 841mm×597mm.

    3 Scanning digitizing software should have the following basic functions:

    1) Drawing orientation and correction.
    2) Data collection and coding input.
    3) The calculation of data, transform, and editing.
    4) Real time graphic display, check and modification.
    5) Drawing the point, line, surface terrain symbol.
    6) The layered management factors of topographic map.
    7) The operation of grid data, including translation, grating image gray value transform and the combination of the grating image, etc.
    8) Coordinate transformation.
    9) Refinement of the linear grid data.
    10) Automatic tracking the grid data of vector quantization.
    11) The man-machine interactive vector quantization.

6.3.2 Scanning the original image should meet the following requirements:

    1 The scale of the original image should not be less than the scale of digital topographic map.

    2 The base drawing with polyester film should be used for the original map. On the premise of meeting users' requirement, other papery graph may be chosen.

    3 Drawings should be level off, without fold, and clear.

    4 Deformation of the original drawings or scan images should be revised.

6.3.3 Scanning digitizing shall meet the following requirements:

1 Drawing and image orientation shall meet the following requirements:

    1) It is advisable to use the quadrangular coordinate point or grid point of the inner vertical-section as the orientation point.

    2) Orientation points should not be less than 4 points, and its location should be distributed evenly and reasonable.

    3) It should be appropriate to increase the paper orientation point when the deformation of the topographic map is large.

    4) The accuracy of grid should be checked after orientation; Directional coordinate with actual coordinate difference allows for 0.3mm on the drawing.

2 The digital topographic map elements shall meet the following requirements:

    1) For drawing control points and building detail points, coordinate input mode should be adopted. The control points without coordinate data can not be drawn.

    2) Border and grid map should adopt the method of the input coordinates automatically generated by the drawing software according to the theoretical value.

    3) When the original topographic map of terrain or object symbol does not accord with the current schema, it should use the currently valid schema rules of symbols.

    4) The point symbols, line symbols, landform and vegetation filling symbols, should be automatically generated by the drawing software.

    5) Digital contour and terrain contour should adopt the

way of line tracking.

3 After each image digitization, the map edge and edge data should be edited. After the edge is completed, the check chart should be output. Check figure compared with the original image, point symbol and flair point deviation should be allowed for the 0.2mm on the drawing, line symbols deviation should be allowed for the 0.3mm on the drawing.

6.4 Topographic Map Revision

6.4.1 Before topographic map revision, field survey should be conducted, the scope of the revision should be determined, and the revision plan should be made. It should be re-measured when the revision area is greater than 20%.

6.4.2 Mapping control of topographic map revision shall meet the following requirements:

1 Appropriate use of the original control points through inspection qualified.

2 When partial revision survey is implemented, the coordinates of station points may be confirmed with known points of original image feature by using the method of interpolation or intersection. And the permissible discrepancy shall be 0.2mm on the map.

3 For the elevation revision survey of a few points in local area, the revision survey may be implemented according to the elevation of 3 fixed detail points, the average value shall be used if the permissible discrepancy of elevation is no more than 1/5 basic distance.

4 If the changed area of plan metric features is bigger, or the round features used as the control are insufficient, some mapping control points shall be set up first.

6.4.3 Topographic map revision shall meet the following requirements:

1 The permissible RMSE shall be 0.6mm in the map between the new features and the original features.

2 The connection part of features revision shall be started from the unchanged position, and a certain amounts of coincident points shall be surveyed in the connection part of landform revision.

3 In addition to the revision for the changed topography, the apparent errors of features and landform in the original topographic map shall be corrected. If errors of the features, landform, annotation and layering in original data appear, correction shall be implemented.

4 When the paper topographic map revision is implemented, it shall be done after digitizing the original map.

5 After the accomplishment of revision survey, the revision condition shall be recorded and sketch shall be drawn according to the mapsheet.

6 The other requirements of the topographic map revision should be carried out in accordance with the provisions of Section 6.2.

6.5 Topographic Map of Check

6.5.1 Topographic map check include: border points, the grid intersection points, mathematical basis inspection of control points, plane position and elevation accuracy check, sheet splicing precision inspection, attribute accuracy check, logical consistency check, finishing quality check, and the attachment quality inspection.

6.5.2 Topographic map inspection shall meet the following requirements:

1 Mathematical foundation check: according to the retrieval condition, the border points, grid intersection point and control point coordinates are displayed on the screen, and then the theoretical value and control point coordinates are checked.

2 Plane position and elevation accuracy examination shall meet the following requirements:

    1) The general provisions of check point selection: The check points of digital topographic map shall be evenly distributed, and randomly selected from apparent feature points. If the total number of mapsheet does not exceed 50, the sampling number shall not be less than 10% of the total number. If the total number of mapsheet is more than 50, the sampling number shall not be less than 5% of the total number. The check points of each sampling mapsheet shall be 20 – 50 points.

    2) Inspection method: the horizontal coordinate and elevation of check points shall be surveyed by using the method of disperse points in field.

3 The detection of the edge accuracy should be done by measuring the distance between the endpoints of the elements at the edge of the two adjacent edges. The deviation values of the unconnected elements are recorded, the edge of the geometry of the natural connection is checked, and the same situation of the field properties and line planning attributes are checked. Record the number of feature entities that do not match attributes.

4 Attribute precision inspection shall meet the following requirements:

    1) Check the correctness of the name of each layer and occurrence of layer, leakage.

    2) Check layers one by one of the correctness and leakage

of attribute items in each attribute table.

3) Check each factor attribute value, code and notes, and its correctness.

4) Check the common edge attribute values and their correctness.

5 Logical consistency inspection shall meet the following requirements:

1) Check whether there is a duplicate element in each layer.

2) Check the direction of the directional symbol and the directional element and its correctness.

3) Check the polygon closure and correctness of identification code.

4) Check the node matching condition of linear elements.

5) Check whether the relationship between elements of expression is reasonable and whether it can properly reflect the distribution characteristics of the elements and density characteristics of each factor, if there are geographical adaptive contradictions.

6) Check whether water system and road elements are continuous.

6 Finishing quality inspection shall meet the following requirements:

1) Whether the elements are correct and the size is in accordance with the scheme rules.

2) Whether the graphics line is continuous, smooth and clear, and whether the thickness is in compliance.

3) Whether the relation of elements is reasonable. Whether there is overlap or gland phenomenon.

4) Check whether the name note is correct, and the location is reasonable. As well as whether the font style,

　　　　font size, font direction are in accordance with the regulations.
　　5) Check whether the note is gland important features or point symbols.
　　6) Check inside and outside surface configuration, and whether border finishing is in line with the regulations.
7 Attachments quality inspection shall include the inspection of correctness and completeness of documents.

6.6　Document and Data Compilation

6.6.1 After completion of digital topographic survey, the following documentation should be integrated:
　　1　The project technical specification.
　　2　Instrument verification and identification documents.
　　3　Measurement record hand book.
　　4　Measuring calculation data and results table.
　　5　Digital topographic map and index map.
　　6　Technical summary report.
　　7　Other documents and materials related to the project.

6.6.2 After the completion of the project, the following documents should be submitted:
　　1　The technical design documents.
　　2　Digital topographic map and index map.
　　3　Technical summary report.

7 Aerial and Space Photogrammetry

7.1 General Requirements

7.1.1 This chapter is applicable to the production of mapping products at scales of 1 : 500 to 1 : 10000 digital line graphs (DLG), digital elevation models (DEM) and digital orthophoto maps (DOM).

7.1.2 The horizontal position permissible RMSE of the photo-control point relative to the closest basic horizontal control point, the vertical position permissible RMSE of the photo-control point relative to the closest basic vertical control point, the horizontal position permissible RMSE and vertical position permissible RMSE of the aerial triangulation point (joint point) obtained from interior work relative to the closest mapping base control point, the horizontal position permissible RMSE and vertical position permissible RMSE of the feature point and elevation point with notes in the DGL relative to the closest mapping base control point, and the vertical permissible RMSE of the contour line in the DLG map sheet shall all be implemented in accordance with the requirements in Table 7.1.2-1 and Table 7.1.2-2.

7.1.3 The DEM data of water and hydropower projects shall be composed of the regular grid point data, feature point data and boundary data. The basic grid size shall be 5m×5m. When mapping the water and hydropower projects junction areas and other buildings in major projects, 2.5m×2.5m, 2m×2m and 1m×1m shall be chosen as the grid sizes. The elevation RMSE of the grid points shall comply with the requirements in Table 7.1.3, and only one decimal place after the unit (meter) is kept for the elevation data.

7.1.4 The horizontal position RMSE of DOM shall be implemented in accordance with the requirements in Table 7.1.2 – 1 and Table 7.1.2 – 2, and the technical indexes such as the ground simpling distance, gray-scale, wave band and edge matching tolerance shall be implemented in accordance with the requirements in Table 7.1.4.

7.1.5 The RMSE of aerial triangulation points (joint points) should be calculated based on the discrepancy of the check points, and the horizontal and vertical RMSE of check points shall be calculated according to Equation (7.1.5 – 1); the horizontal and vertical RMSE of common points among block of flight strips shall be calculated according to Equation (7.1.5 – 2), and the requirements of its accuracy shall be the same as those for the joint points.

$$m_C = \pm \sqrt{\frac{[\Delta \Delta]}{n}} \qquad (7.1.5-1)$$

$$m_N = \pm \sqrt{\frac{[dd]}{3n}} \qquad (7.1.5-2)$$

Where:

m_C = the RMSE of a check point, m;

m_N = the RMSE of a common point, m;

Δ = the discrepancy between the field observed value and the calculated value of a check point, m;

d = the discrepancy of a common point among block of flight strips, m;

n = the number of points used to access the evaluation precision.

7.1.6 The vertical RMSE of DLG in large stretches of forests, marshes and desert areas may be extended to 1.5 times the specified values in Table 7.1.2 − 1 and Table 7.1.2 − 2; the vertical RMSE of DEM may be extended to 1.5 times the specified values in Table 7.1.3 and the vertical accuracy of DEM interpolation points may be extended to 1.2 times the vertical accuracy of grid points; the horizontal position permissible RMSE of DLG and DOM may be extended to 1.5 times the RMSE, but the error shall not be greater than 1.0mm in the mountains and high mountains maps. Other special needs shall be specified definitely in the technical design.

7.1.7 The methods of digital aerophotogrammetrical mapping should be selected in accordance with the mapping methods in Table 7.1.7 according to the aerophotogrammetrical data and mapping accuracy. In special circumstances, other aerophotogrammetrical mapping methods can also be used, but the mapping accuracy shall meet the accuracy requirements specified in Table 7.1.2 − 1 and Table 7.1.2 − 2.

7.1.8 The aerophotogrammetrical data shall meet the following requirements:

1 The mapping scale of aerial photography shall be selected according to actual requirements of the projects; and considering the current aerophotogrammetrical techniques, equipment and the situation of project areas, the aerophotogrammetrical scale with good quality and reasonable cost, image ground sampling distance, flight height, focal length and type of aerial photographic camera, GNSS-assisted aerophotography and IMU/DGNSS-assisted aerophotography shall be selected, the aerophotography seasons shall be selected too, and other special technique requirements shall be proposed.

2 The aerophotogrammetrical scale of aerial photography

Table 7.1.2 – 1 Stipulation of mapping accuracy of scale 1 : 500, 1 : 1000, and 1 : 2000 by using aerial photogrammetry

Mapping scale	1 : 500				1 : 1000				1 : 2000			
Code of data	DLG500				DLG1000				DLG2000			
Topographic category	Plains	Hills	Mountains	High mountains	Plains	Hills	Mountains	High mountains	Plains	Hills	Mountains	High mountains
Basic contour interval (m)	0.5	0.5	1.0	1.0	0.5	0.5	1.0	1.0	0.5	1.0	1.0	2.0	2.0
Photo-control points — Horizontal position RMSE (mm in the map)	0.10				0.10				0.10			
Photo-control points — Vertical RMSE (m)	0.05	0.05	0.10	0.10	0.10	0.05	0.05	0.10	0.10	0.20	0.05	0.10	0.10	0.20	0.20
Interior work pass points — Horizontal position RMSE (mm in the map)	0.40			0.55	0.40			0.55	0.40			0.55
Interior work pass points — Vertical RMSE (m)	—	0.30	0.45	0.65	—	0.23	0.30	0.45	0.65	1.30	—	0.23	0.30	0.45	0.90	1.30

Table 7.1.2 – 1 (Continued)

Mapping scale	1:500				1:1000				1:2000			
Code of data	DLG500				DLG1000				DLG2000			
Topographic category	Plains	Hills	Mountains	High mountains	Plains	Hills	Mountains	High mountains	Plains	Hills	Mountains	High mountains
Basic contour interval (m)	0.5	0.5 1.0	1.0	1.0	0.5	0.5 1.0	1.0	2.0	0.5	1.0	1.0 2.0	2.0
Horizontal position RMSE of features (mm in the map)	0.6		0.8		0.6		0.8		0.6		0.8	
Vertical RMSE of elevation points (m)	0.14 0.20 0.40		0.60	0.80	0.14 0.30 0.40		0.60	0.80 1.60	0.14 0.30 0.40 0.60		1.20	1.60
Vertical RMSE of map-sheet contour lines (m)	0.17 0.25 0.50		0.70	1.00 point of vertical curve	0.17 0.35 0.25 0.50		0.70	1.00 2.0 point of vertical curve	0.17 0.35 0.50 0.70		1.40 point of vertical curve	2.00 point of vertical curve

Note: Full field horizontal and vertical control point distribution method shall be adopted when horizontal and vertical position RMSE of pass points in interior work cannot meet the accuracy requirements in the table. The condition shall be suitable for use only in mapping 1:500 plains and hills and mapping 1:1000 and 1:2000 plains.

Table 7.1.2 – 2   Stipulation of mapping accuracy of scale 1 : 5000 and 1 : 10000 by using aerial photogrammetry

Mapping scale		1 : 5000				1 : 10000			
Code of data		DLG5000				DLG10000			
Topographic pattern		Plains	Hills	Mountains	High mountains	Plains	Hills	Mountains	High mountains
Basic contour interval (m)		0.5	1.0	2.0	5.0	0.5	1.0	2.5	5.0 (10.0)
Photo-control points	Horizontal position RMSE (mm in the map)	0.10				0.10			
	Vertical RMSE (m)	0.05	0.10	0.20	0.50	0.05	0.10	0.25	0.50
Interior work pass points	Horizontal position RMSE (mm in the map)	0.35			0.50	0.35			0.50
	Vertical RMSE (m)	—	0.23	0.30	0.60	0.90	2.00	3.00	—	0.23	0.30	0.60	0.80	2.00	3.00
Horizontal position RMSE of features (mm in the map)		0.50			0.75	0.50			0.75
Vertical RMSE of elevation points (m)		0.14	0.30	0.40	0.80	1.20	2.50	4.00	0.14	0.30	0.40	0.80	1.00	2.50	4.00
Vertical RMSE of map-sheet contour lines (m)		0.17	0.35	0.5	1.0	1.40	3.00 point of vertical curve	5.00 point of vertical curve	0.17	0.35	0.5	1.0	1.25	3.00 point of vertical curve	5.00 point of vertical curve

Notes: 1. Full field horizontal and vertical control point distribution shall be used when vertical RMSE of pass points in interior work for plains could not meet the accuracy requirement in table.
2. When mapping the 1 : 10000 scale map of high mountainous, contour interval of 10m shall be used when the angle of inclination is larger than 40° in most of land within tables; The contour interval of 5 or 10m shall be used according to the project situation when the angle of inclination is no greater than 40° in most of land within map.
3. When mapping the 1 : 10000 scale map, contour interval of 0.5m shall be allowed for use only in case the angle of inclination of land is less than 1° and when necessary.

surveying instrument with conventional photographic film or the image ground sample distance of aerial photography surveying with digital aero surveying instruments should be chosen according to the range listed in Table 7.1.8. The selection shall be based on the topography of project areas, mapping accuracy requirements, and the performance, image-forming principle, base-height ratio, photogrammetric distortions, imaging quality of the aerophotogrammetrical camera, and the vertical accuracy of the selected aerophotogrammetrical scale denominator M and image ground sample distance (GSD) shall be estimated according to Equation (7.1.8-1). The M and GSD with a smaller k shall be chosen if the requirements of mapping accuracy are high and the performance and technical indexes of the aerophotogrammetrical cameras are low. Because of the limitation of various conditions, when the M and GSD with a larger k are chosen, the configuration of full field photo control point distribution, laying out the artificial ground target and other technique measures shall be adopted.

Table 7.1.3 Technical indexes of DEM

Code of data	Grid size	Accuracy grade	Vertical RMSE of grid points (m)			
			Plains	Hills	Mountains	High mountains
DEM-SLA1	5m	A	0.35	1.25	3.00	5.00
DEM-SLA2		B	0.48	1.69	4.10	6.80
DEM-SLA3		C	0.70	2.50	6.00	10.00
DEM-SLB1	2.5m/ 2m/1m	A	0.17	0.50	1.40	2.00
DEM-SLB2		B	0.23	0.68	1.90	2.70
DEM-SLB3		C	0.35	1.00	2.80	4.00

Table 7.1.4 Technical indexes of DOM

Scale		1:500	1:1000	1:2000	1:5000	1:10000
Codes of data		DOM500	DOM1000	DOM2000	DOM5000	DOM10000
Ground simpling distance of images (m)		0.05	0.1	0.2	0.5	1.0
Gray scale (radiation resolution)		256 gray scale				
Wave band		One or more (such as RGB)				
Tolerance of edge matching (m)	Plains and hills	0.3	0.6	1.2	2.5	5
	Mountains and high mountains	0.4	0.8	1.6	3.75	7.5

Table 7.1.7 Methods of aerial photogrammetry mapping

Contour interval	Ground sample distance of images (GSD)	Methods of aerial photogrammetry mapping
0.5m	(1) GSD ≤ denominator of mapping scale (1:500) ×0.01cm; (2) GSD ≤ denominator of mapping scale (1:1000 and 1:2000) ×0.008cm; (3) GSD ≤ denominator of mapping scale (1:5000 and 1:10000) ×0.006cm	1. Control points of photograph shall be laid out in horizontal block of flight strips; digital aerotriangulation shall be densified in horizontal; the total digital mapping workstation shall be used in surveying ground objects and making DOM; the elevation points with notes shall be surveyed in the field and the contour lines shall be drawn. 2. Horizontal and vertical control points shall be laid out by using full field distribution; the total factor method by using single model orientation of total digital mapping workstation shall be used in mapping and making DEM and DOM. 3. Horizontal photo control points shall be laid out in the block of flight strips; vertical control points shall be laid out by using full field distribution; digital aerotriangulation shall be densified in horizontal; the total factor method of total digital mapping workstation shall be used in mapping and making DEM and DOM.

Table 7.1.7 (Continued)

Contour interval	Ground simpling distance of images (GSD)	Methods of aerial photogrammetry mapping
0.5m	GSD is higher than the value specified in the above requirements.	4. Method 2 mentioned above shall be adopted; 6 control points of photograph shall be laid out in each image pair; the four corners of the mapping scope and principal point of photograph shall lay out a control point respectively in each image pair. 5. The surveying marks on land shall be laid out, and Methods 1, 2 and 3 shall be adopted.
1m	(1) GSD≤denominator of mapping scale (1:1000) × 0.01cm; (2) GSD≤denominator of mapping scale (1:2000) × 0.008cm; (3) GSD≤denominator of mapping scale (1:5000 and 1:10000) ×0.006cm	6. Control points of photograph shall be laid out in horizontal and vertical block of flight strips; digital aerotriangulation shall be densified in horizontal and vertical; the total factor method of total digital mapping workstation shall be used in mapping and making DEM and DOM. 7. The high-accuracy IMU/DGNSS shall be auxiliary for the aerophotography; control points of photograph shall be laid out in horizontal and vertical block of flight strips; digital aerotriangulation shall be densified in horizontal and vertical; the total factor method of total digital mapping workstation shall be used in mapping and making DEM and DOM. 8. When the push-broom digital aero photographical camera is working, a few horizontal and vertical control points of photograph and check points shall be laid out in field; digital aerotriangulation shall be densified in horizontal and vertical; the total factor method of total digital mapping workstation shall be used in mapping and making DEM and DOM.
	GSD is higher than the value specified in the above requirements	9. Methods 1, 2, 3 and 4 shall be adopted. 10. The surveying marks on land shall be laid out, and Methods 6, 7 and 8 shall be adopted.

Table 7.1.7 (Continued)

Contour interval	Ground simpling distance of images (GSD)	Methods of aerial photogrammetry mapping
2-5 (10) m	(1) GSD⩽denominator of mapping scale (1:1000) × 0.012cm; (2) GSD⩽denominator of mapping scale (1:2000) × 0.01cm; (3) GSD⩽denominator of mapping scale (1:5000 and 1:10000) ×0.008cm	11. Method 6 above shall be adopted. 12. The high-accuracy IMU/DGNSS shall be auxiliary for the aerophotography; horizontal and vertical control points of photograph shall be laid out for block of flight strips according to the regulation of "four corners and two sides" or "four corners and two lines"; the digital aerotriangulation shall be densified in horizontal and vertical; the total factor method of total digital mapping workstation shall be used in mapping and making DEM and DOM. 13. When the push-broom digital aerial photographic camera is working, check points shall be laid out in the field work; the digital aerotriangulation shall be densified in horizontal and vertical; the total factor method of total digital mapping workstation shall be used in mapping and making DEM and DOM.
	GSD is higher than the value specified above requirements	14. Methods 1, 2, 3 and 4 mentioned above shall be adopted. 15. The surveying marks on land shall be laid out, and Methods 6, 7 and 8 shall be adopted.

Note: For the aerial photography by using the conventional film camera, GSD = scanning resolution of image×denominator of scale of aerial photograph.

Table 7.1.8 Selection range of aerial photographic scale or image ground sample distance

Mapping scale	Denominator of aerial photographic scale of conventional film camera M	Image ground sample distance of digital aerial surveying camera (GSD) (cm)	k
1:500	2000⩽M⩽4000	4⩽GSD⩽8	4.0-8.0
1:1000	3000⩽M⩽7000	6⩽GSD⩽14	3.0-7.0

Table 7.1.8 (Continued)

Mapping scale	Denominator of aerial photographic scale of conventional film camera M	Image ground sample distance of digital aerial surveying camera (GSD) (cm)	k
1:2000	$4000 \leqslant M \leqslant 12000$	$8 \leqslant GSD \leqslant 24$	2.0 – 6.0
1:5000	$7000 \leqslant M \leqslant 25000$	$14 \leqslant GSD \leqslant 50$	1.4 – 5.0
1:10000	$10000 \leqslant M \leqslant 40000$	$20 \leqslant GSD \leqslant 80$	1.0 – 4.0

Notes: 1. For the aerial photography by using the conventional film camera, k is the ratio of photo the scale denominator to the mapping scale denominator; for the aerial photography by using the digital camera, k is the ratio of the aerial photograph scale denominator by using equivalent conventional film camera to the mapping scale. The denominator of aerial photograph scale by using equivalent conventional film camera shall be calculated as denominator of digital aerial photographic scale $\times$ pixel size/20 or GSD$\times$10000/20.

2. For the aerial photographic scale by using the conventional film camera, the average relative flying height (H, unit of measurement is m) and focal length of aerial photographic camera (f, unit of measurement is mm) should be determined by equation $M = H/f \times 1000$ according to the topographic situation, and resolution of aerial photograph scanning and achievable vertical accuracy shall be estimated in accordance with Equation (7.5.1-1) and Equation (7.1.8-1).

3. For the ground sample distance of image selected by aerial photography by using the digital camera, the scale denominator of aerial photography and relative flying height (H, its measurement unit is m) should be calculated and determined by the equation $M = GSD/P_i \times 10000$ and $H = Mf/1000$ in accordance with resolution (P_i is the size of pixel, and the unit of measurement is μm) and focal length (f, its measurement unit is mm) of digital camera which is planned to adopt, and the achievable vertical accuracy shall be estimated in accordance with Equation (7.1.8-1).

3 The achievable vertical accuracy of mapping in accordance with the selected aerial photography data may be estimated by using Equation (7.1.8-1).

$$\Delta h = \frac{1.25 R_P H}{1000 b} \qquad (7.1.8-1)$$

Where:

Δh = vertical accuracy (the limit of RMSE of height of pass points) that is required, m;

H = average relative flying height, m;

b = average length of photo base line, mm; For the average length of base line of digital aerial photographic image, the calculation shall be carried out in accordance with the Equation (7.2.3-3);

R_P = scanning resolution of image, μm, rounding off numbers. For the digital aerial photographic image, $R_P = P_i$;

P_i = size of pixel, μm.

4 The below requirements of aerial photography shall be met:

1) The design drawing for the aerial photography plan: if the scale of aerial photography is no less than 1 : 3500, the topographic map on a scale of 1 : 10000 should be adopted; if the scale of aerial photography is no more than 1 : 10000, the topographic map on a scale of 1 : 50000 or 1 : 25000 should be adopted; if the scale of aerial photography is no more than 1 : 10000, the topographic map on a scale of 1 : 50000 or 1 : 25000 should be adopted; or the processes shall be proceeded in the 3D building-model system.

2) Principles which shall be followed in the division of the aerial photography blocks: the boundary between aerial photography blocks shall be consistent with the mapping sheet line. If the scale of aerial photography is less than 1 : 7000, the topographic height difference in the block shall not be greater than 1/4 of the

relative flying height; if the scale of aerial photography is no less than 1 : 7000, the topographic height difference in the block shall not be greater than 1/6 of the relative flying height; if it is permitted by the topographic conditions, the larger block of aerial photography should be divided. If the topographic height difference can meet the requirements and the direct linear of air routes can be guaranteed, larger spacing of block should be divided.

3) The height of block datum plane shall be calculated in accordance with Equation (7.1.8-2) below:

$$h_{base} = \frac{1}{2}(h_{averageH} + h_{averageL}) \quad (7.1.8-2)$$

$$h_{averageH} = \frac{1}{n}\sum_{i=1}^{n} h_{iH} \quad (7.1.8-3)$$

$$h_{averageL} = \frac{1}{n}\sum_{i=1}^{n} h_{iL} \quad (7.1.8-4)$$

Where:

h_{iH}, h_{iL} = the height of the typical high points and low points in the block, m;

n = the number of the elevations of the high points or low points.

4) The following principles shall be followed in laying out the air route: the flight direction of the air route should fly in an east-west rectilinear line. In the particular situation, the direction may be rectilinear course in an arbitrary direction such as south-north or along with a wire, a river, a coast and a boundary. Exposure points should be designed by using DEM in a

point-by-point way according to the topographic fluctuation. When a water area or a sea area is shot, the principal point of the photograph should not be placed in water. Every island should be covered totally and the normal overlapping stereopair shall be built.

5) The principles which shall be followed when season and time of aerial photography are chosen: the most favorable meteorological condition in the project areas shall be chosen, and the negative influences of aerial photography caused by land vegetation and other coverings should be avoided or reduced, and the terrain details can be showed truly by using the image of the aerial photography; in the process of the aerial photography, the sufficient illuminance shall be guaranteed and oversize shadow shall be avoided. The season and time of aerial photography should be determined by solar elevation and multiple of object shadow in the project area of aerial photography. The solar elevation shall be greater than 20° in plains while the enlarged multiple of the object shadow shall be less than 3. The solar elevation shall be greater than 30° in hills and the area of small towns while the enlarged multiple of the object shadow shall be less than 2. The solar elevation shall be greater than 45° in mountain land and the area of medium-sized cities while the enlarged multiple of the object shadow shall be less than 1. In the cliffy mountain area and dense high-rise buildings with extraordinary large elevation difference, the time of aerial photography shall be limited in 1h around the noon and the enlarged multiple of the object shadow shall be

less than 1, and the aerial photography below clouds could be implemented if it could be done in the condition; if the flight areas are in desert, gobi, forest, grassland, widespread salt flat and saline-alkali soil, the aerial photography shall not implemented in 2h around the noon.

6) Requirements of flight quality: longitudinal overlap should be 60%-65%; The largest individual shall not be greater than 75% while the minimum shall not be less than 56%; degree of lateral overlap should be 30%-35%, and the minimum shall not be less than 13% while the indicator may be extended appropriately in accordance with national standards in particular cases. Tilt angle of photograph should not be greater than 2°and the largest individual shall not be greater than 4°. If the scale of aerial photography is less than 1 : 7000 and relative flying height is greater than 1200m, swing angle of aerial photographic should not be greater than 6°, and the largest individual should not be greater than 8°. If the scale of aerial photography is less than 1 : 3500 but greater than or equal to 1 : 7000, swing angle of aerial photographic should not be greater than 8°, and the largest individual should not be greater than 10°. If the scale of aerial photography is no less than 1 : 3500, swing angle of aerial photographic should not be greater than 10°, and the largest individual should not be greater than 12°. When the digital mapping method is adopted in particular cases, swing angle of aerial photographic may be extended by 2° in accordance with the relevant

provisions under the premise of flight course of aerial photograph and degree of lateral overlap meeting the requirements; in the same flight course, the number of aerial photograph with swing angle reaching or approaching the largest swing angle shall not be greater than 3 pieces. If the digital mapping method is adopted in particular cases, swing angle of photograph may be extended by 2° in accordance with the relevant provisions, and the fight course of photograph and degree of lateral overlap shall meet the requirements. In the same flight course, the number of aerial photographs with the swing angle reached or very close to the maximum swing angle shall not be greater than 3. In the same working area, the number of the aerial photographs with the maximum swing angle shall not be greater than 4% of the total number of the aerial photographs, and the strip deformation shall not be greater than 3%. In the same flight course, the difference between the flight heights of two neighboring photographs shall be not greater than 20m, and the difference between the maximal flight height and the minimal flight height shall not be greater than 30m. The difference between the actual flight height and the design flight height shall not be greater than 50m in the block of aerial photography. When relative flight height is greater than 1000m, the allowable difference between the actual flight height and the design flight height shall be 5% of the design flight height.

7) The coverage requirements for the flight area and flight block of aerial photography: the part of longitu-

dinal overlap exceeding the working area border side shall not be less than one baseline. The part of lateral overlap exceeding the working area border side shall not be less than 50% of the picture format. When the measurement of photo control points and regular densification in indoor work are not influenced, the part of lateral overlap exceeding the working area border side may be extended to no less than 30% of the picture format.

  8) Requirements of repairing leaks: the relative and absolute leaks in aerial photography shall be repaired in time by using the digital camera which is used in the last aerial photography, and the both ends of repairing course shall exceed a photographic baseline.

5 If GNSS or IMU/DGNSS is used as an auxiliary in the aerophotography, the layout and survey of the ground base station and selection and survey of calibration field shall be carried out, and the layout, selection, observation and data processing shall be implemented in accordance with the current national relevant standards.

6 The other requirements and acceptance checks for the aerial photography and the aerial photography data may be implemented in accordance with the national aerial photography standards of the relevant mapping scale.

7.1.9 The data of space photography shall meet the following requirements:

  1 The ground sample distance of the satellite image shall be selected and used in accordance with Table 7.1.9 below.

  2 The data of satellite images and the collection of relative data shall meet the requirements below:

Table 7.1.9 Selection range of ground sample distance of satellite images

Mapping scale	Ground sample distance of images (m)
1 : 5000	≤0.6
1 : 10000	≤1.0

1) Panchromatic and multi-spectral satellite stereo photos with relatively new time phase shall be selected, and the acquisition time of image should be consistent or similar. And the precise parameters such as ephemeris and the angle of attitude related to the image shall be collected.

2) The collected data shall meet the requirements of DLG production.

7.2 Photo Control Point Design

7.2.1 The position of photo control point shall meet the following requirements:

1 The target image of photo control point position should be clear, easy to distinguish and stereo-measure. If the contradictions occur between the target image and other image conditions, the target image conditions shall be taken into consideration emphatically.

2 The photo control point should be used for public. For the frame perspective image, it should be set in the flight course longitudinal overlap within 3 pieces; if there is a public use of the up and down adjacent flight course, it shall be set in longitudinal overlap and lateral overlap of flight course within 6 or 5 pieces.

3 The distance from the photo control point to the photo edge shall not be less than 1.5cm (picture format 23cm×23cm) or 5% of the picture format width, and the distance from the photo control point to various signs on the photo shall not be less than 0.1cm or 50 pixels; for images obtained by the digital aerial camera, the distance from the photo control point to the photo edge should not be less than 0.1cm or 50 pixels.

4 The photo control point shall be laid out near the mid line of the lateral overlap, the distance leaving from the bearing line shall be longer than 4cm (picture format 23cm×23cm) or 20% of the picture format width; when the lateral overlap is too large to satisfy the requirement of the distance leaving from the bearing line, the points shall be set respectively; when the lateral overlap is excessively little and the photo control point on up and down adjacent routes could not be used in common, the points should be laid out respectively, the vertical distance between two splitting points should not be longer than 1cm, and the maximum should not be greater than 2cm.

5 The photo control point position of the space remote sensing imagery shall be easy to be measure exactly, and the distance leaving from photo edge is not restricted if the observation is not influenced.

6 If the survey area is divided according to the table line, the point located in the free table margin, uncompleted table margin and other table margins shall be laid out outside the table line. When a survey area is divided according to the mapping needs, control points should be laid out outside the scope line of mapping needs.

7.2.2 The layout of total field photo control point shall meet the following requirements:

1 When the full field control point distribution of image maps made by single photo digital rectification is implemented, if the mapping scale is not greater than 4 times the aerophotography scale, a horizontal and vertical point shall be laid out in each corner on the spaced image region of surveying and mapping, and a horizontal and vertical check point should be set near the principal point. If the mapping scale is greater than 4 times the aerophotography scale, more control points shall be laid out.

2 The stereoscopic total factor mapping requires that a horizontal and vertical control point should be laid out in each of 4 corners of the surveying and mapping region of every stereoscopic image pair. If the mapping scale is larger than 4 times the aerophotography scale, one more horizontal and vertical control point shall be laid out near the principal point of photo.

3 The photo control point position of the total field on the photo shall meet the requirement of Article 7.2.1, and the control points should leave from the principal point of photo, and the distance perpendicular to the bearing line should be less than 5% of the picture format width. Several points shall not be longer than 8% of the picture format width in difficult conditions.

7.2.3 The layout of block of flight strips photo control points shall meet the following requirements:

1 The division of block of flight strips should be based on aerial partition, flight course, distribution of map table, terrain conditions, etc. The shape of network should be a rectangular. The size of network and span between control points shall be determined in accordance with the factors such as mapping precision, aerial data condition and the process of the system error. The division of block of flight strips shall meet the following requirements:

  1) When block of flight strips is divided, the situation of jointing flight courses on first and last course of block

of flight strips should be avoided.

2) The block of flight strips shall not include the courses and image pairs with photographic overlap that do not meet the requirements, and shall not include the image pairs with large cloud and shadow, etc., which would affect the connection of densified control network in interior work.

3) When different map sheet are divided in a same network, the layout of points shall meet the requirement of higher precision relatively.

4) If the photos made by the same photographic scale and different aerial camera are divided in the same network, an additional horizontal and vertical control point in the course connection of each flight course shall be laid out.

5) There shall be a horizontal and vertical control point in the center of block of flight strips for examination.

2 The number of routes between two adjacent planar points in block of flight strips, shall be set as shown in Table 7.2.3 - 1.

Table 7.2.3 - 1 **The number of interval routes stipulated by k**

K	$k<2$	$2 \leqslant k<4$	$k \geqslant 4$
The number of interval routes (item)	$\leqslant 8$	$\leqslant 6$	$\leqslant 4$

3 The baseline number n of block of flight strips along the course between adjacent horizontal points and vertical points, shall not be greater than the n value calculated in Equation (7.2.3 - 1) and Equation (7.2.3 - 2) respectively. When a conflict occurs between the permissible horizontal and the vertical baseline number, the vertical number shall be the first priority. When the aerial photography is implemented without the as-

sistant of the GNSS aided aerial photography and IMU/DGNSS, n shall not be greater than 12.

$$m_s = \pm 0.28 \times \frac{H}{f} \times \frac{1}{M} \times m_q \times \sqrt{n^3 + 2n + 46}$$
(7.2.3-1)

$$m_h = \pm 0.088 \times \frac{H}{b} \times m_q \times \sqrt{n^3 + 23n + 100}$$
(7.2.3-2)

$$b = L_x \times (1 - P_x) \times P_i / 100 \quad (7.2.3-3)$$

Where:

m_s = horizontal RMSE of pass points, mm in the map;
m_h = vertical RMSE of pass points, m;
M = denominator of the mapping scale;
m_q = RMSE with unit weight of parallax measurement, mm;
n = number of baselines between the adjacent control point;
H = relative flying height, m;
f = focal length of aerial camera, m;
b = average length of photo baseline, mm. For the average baseline length of digital aerial images, the calculation shall be implemented according to Equation (7.2.3-3);
L_x = course breadth of digital images, pixel;
P_x = average course overlap degree of digital images, %;
P_i = pixel size or image resolution, μm.

4 When n among the horizontal and vertical points are calculated and the horizontal RMSE m_s of pass points and the vertical RMSE m_h of pass points are estimated, the m_q in Table 7.2.3-2 and Table 7.2.3-3 shall be used.

5 The control point distribution for block aerotriangulation scheme may be implemented in accordance with the following requirements, or may be implemented with GB/T 7931 and GB/T 13977.

Table 7.2.3-2 m_q shall be adopted in calculating the baseline number n among horizontal points or estimating the horizontal RMSE m_s of pass points

	Item	$n \leqslant 4b$		$n > 4b$	
		Do not set ground symbol	Set ground symbol	Do not set ground symbol	Set ground symbol
m_q	Plains and hills	$1.5P_i/1000$	$P_i/1000$	$1.1P_i/1000$	$0.75P_i/1000$
	Mountains and high mountains	$1.75P_i/1000$	$1.25P_i/1000$	$1.35P_i/1000$	$P_i/1000$

Table 7.2.3-3 m_q shall be adopted in calculating the baseline number n among vertical points or estimating the vertical RMSE m_s of pass points

	Item	Do not set ground symbol	Set ground symbol
m_q	Plains and hills	$1.1P_i/1000$	$0.86P_i/1000$
	Mountains and high mountains	$1.35P_i/1000$	$1.1P_i/1000$

1) After the number of routes in the block of flight strips, the number of routes between two adjacent horizontal points, the number of baselines along strips between adjacent horizontal and vertical points are determined according to the factors of aerial partition, routes, map sheet distribution, topographical condition, precision requirements, etc., one of the schemes in Figure 7.2.3-1, Figure 7.2.3-2 and Figure 7.2.3-3 shall be chosen to lay out points in the block of flight strips. In the graph, "⊙" is the horizontal and vertical point, "•" is the vertical point.

2) When the boundary of the block of flight strips is irregular, irregular control point distribution for block aerotriangulation may be adopted: horizontal and vertical points at the concave and vertical points at the convex should be laid out. When the distance of point of concave angle and point of convex angle is greater than two baselines, the horizontal and vertical points shall be laid out at the concave angle as shown in Figure 7.2.3 - 4.

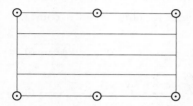

Figure 7.2.3 - 1 Methods 1 of control point distribution for block aerotriangulation

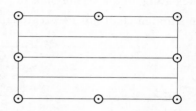

Figure 7.2.3 - 2 Methods 2 of control point distribution for block aerotriangulation

6 Horizontal and vertical control points shall be laid out in block of flight strips for examination.

7 Photo control points of free sheet margin shall be set on sheet margin; it allows that the baseline span outside the photo control points can be 0.7 times normal span.

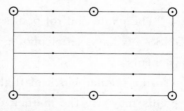

Figure 7.2.3 - 3 Method 3 of control point distribution for block aerotriangulation

In addition to the first and end points on the route, photo control points around the zone should be set on the area of three pieces overlap.

8 When the single route is densified, the two up and down control points at first and end of route may be arranged on the line through the main point and perpendicular to the bearing line, and the deviation from each other should not be greater than half of baseline in the difficult times. The up and down points should be set on the same stereopair; the two control points in the route should be set on the center line between the first and end control points; in difficult situation control points may be deviated to two sides, and one may be on the center line; deviating to the same side of center line shall be avoided, the maximum shall not be greater than a baseline when it comes.

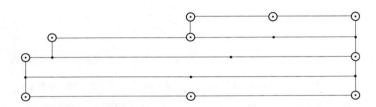

Figure 7.2.3 - 4 Method 4 of control point distribution
for block aerotriangulation

7.2.4 The photo control points layout of block of flight strips by GNSS auxiliary aerial photography shall meet the following requirements:

  1 In a regular block of flight strips, layout of photo control points may use the method of "four corners two sides" or "four corners two lines" as shown in Figure 7.2.4 - 1 and Figure 7.2.4 - 2. The sizes of block of flight strips shall meet the demand of densified precision. The baseline number of course span should be calculated as $\sqrt{3}$ times the n in Equation (7.2.3 - 1), and n shall not be greater than 18; lateral span shall be implemented by the provisions in Table 7.2.3 - 1. In plains, according to the requirement of elevation precise, it is required to add 1 row of vertical control points at intervals of 6 – 8 baselines, or to add 1 row vertical control points to the baseline number calculated as $\sqrt{3}$ times the n in Equation (7.2.3 – 2); 1 or more horizontal and vertical check points in the block of flight strips shall be laid out.

Figure 7.2.4 – 1　"Four corners two sides" method　　　Figure 7.2.4 – 2　"Four corners two lines" method

2　In irregular block of flight strips, additional photo control points shall be set around it. The horizontal and vertical control points should be set at the turning place of a convex angle; the vertical control points shall be laid out when there is 1 baseline at the turning place of concave angle, and horizontal and vertical control points shall be laid out when there are greater than 1 baseline as shown in Figure 7.2.4 – 3.

7.2.5　The photo control point layout of block of flight strips by inertial navigation and differential positioning technology (IMU/DGNSS) auxiliary aerial photography shall meet the following requirements:

　　1　"4 corners" method may be used to lay out photo control points in a regular block of flight strips. A horizontal and

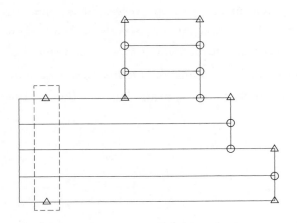

Figure 7.2.4-3 Irregular densified partition layout method

△—Horizontal and vertical control point   ▭—Mapping route
○—Vertical control point Control route   ▭—Control route

vertical control point shall be set respectively at the four corners of block of flight strips, and greater than one horizontal and vertical check point shall be laid out on the center of block of flight strips. The control points of 4 corners should be set in the area within the scope of the edges and two routes overlap, and the center point shall be set on the center of block of flight strips.

2 Size of block of flight strips shall meet the demand of densified precision. For the frame images, spacing along strip shall be calculated as 2 times the n in Equation (7.2.3-1), and the maximum of baselines shall not be greater than 20; spacing across strip shall be implemented in accordance with the regulation of Equation (7.2.3-2). In flatterrain, according to the requirements, it is required to add 1 row of vertical control points at intervals of 6-10 baselines, or to add 1 row of vertical control points to the baseline number calculated as 2 times the n in Equation (7.2.3-2). Spacing across strip shall be taken into consideration only for the digital aerial image of pushbroom.

3 In irregular block of flight strips, photo control points should be set around it, and the horizontal and vertical control points should be set at the flexure of convex angle.

4 For the digital aerial image of pushbroom, one photo control point may be set respectively at the four corners; with the condition of quasi-geoid result, the photo control points may not be laid out, and unconstrained block adjustment shall be used, but greater than 2 horizontal and vertical check points in the block of flight strips shall be laid out.

7.2.6 The layout of photo control points under special condition shall meet the following requirements:

1 In the navigating zone or the joint of aerial partition, the control points shall be set on the joint of routes overlap, and the adjacent regions shall be used in common. If it cannot meet the requirement of use in common, the points shall be set respectively.

2 When the longitudinal overlap degree is less than 53% and aerial photographic gaps exist, the points should be set respectively, and field mapping method shall be used to survey the gaps.

3 When the lateral overlap degree is less than 15%, points shall be set respectively. When the overlap is greater than 1cm and the image is clear and there are no important ground objects within the region, 2-3 elevation points may be added to the overlap area. If the overlap area is less than 1cm or the image is clear, field mapping method shall be used to survey the gaps.

4 When the principal point or standard point is located in the water area, or covered by cloud shadow, shade, snow shadow, etc., or there are no obvious ground objects, the point locations are all considered as points in water. When the size and location of points in water do not affect the stereo model jointing, points may be set as the normal routes. When there is no obvious object within the scope of 2cm leaving from the principal point, or there is no joint point in the region of course three overlap, the image pair in water shall be laid out in accordance with full field control point distribution. The vicinity of standard position of the orientation point is the water zone. If the joint cannot be chosen in the region leaving from 4cm of 23cm ×23cm picture format bearing line or the 15% course three overlap of the picture format, the image pair in water shall be laid out in accordance with full field control point distribution.

 5 In waterfront and island area, points should be set in accordance with full field control point distribution in principle of controlling the mapping area with maximum limit. The land area overrunning 1cm ligature of the control point should add the horizontal and vertical points; in difficult conditions, the elevation points may be used. If the aerial survey method could not guarantee the precision, field mapping method shall be used to implement the additional survey.

7.2.7 The photo control point layout of space remote sensing image shall meet the following requirements:

 1 When mapping is implemented according to the space remote sensing stereopsis by providing parameter of rational polynomial model (RPC), every image pair shall be set at least 3 horizontal and vertical points as photo control points; the photo control points shall not be set along the line, and there should be some distance between the points. When the number of control points is greater than 3, photo control points shall be set evenly in the range of image, and the appropriate number of points shall be chosen as checkpoint.

    2 When several image pairs are mapped in a work area, and there is overlap between the image pairs, photo control points shall use the block of flight strips layout method. A horizontal and vertical point in each of the four corners as the photo control points shall be laid out, and the appropriate number of check points in the block of flight strips shall be set. If the region is large, in addition to setting points at the four corners, photo control points shall be added to the center of the region, and should be set in the region of overlap. The point position can be adjusted if layout of points is difficult. But the quadrilateral or polygon composed of photo control points shall meet the requirement of covering the range of mapping.

    3 When the orthophoto map is produced by single space remote sensing image, if polynomial correction has been used, the photo control points shall not be less than 9, and the points shall be set evenly in the range of image.

7.3 Measurement of Photo Control Points

7.3.1 The precision of photo control points shall be implemented in accordance with the stipulation of Table 7.1.2-1 or Table 7.1.2-2; the measurements of photo control points should be implemented in accordance with requirements of mapping base points, and piling or marking in the field of photo control points shall be implemented.

7.3.2 The interpretation and pricking points of photo control points shall meet the following requirements:

    1 The interpretation and pricking points of photo control points shall be implemented in accordance with the requirements of GB/T 7931 and GB/T 13977.

2 The interpretation and pricking points of photo control points should be photographed by digital camera.

3 For the digital aerial photography, the location of photo control points with the corresponding instructions should be indicated on the image.

7.3.3 The grooming of pricking points for the field photo control points shall meet the following requirements:

1 The print photos, spreadsheet and digital aerial photography can be used to groom pricking points for the photo control points.

2 The grooming of pricking points for the photo control points shall be specified, neat and clear, and instructions shall be concise and accurate; at last, point site, instructions, and punctures shall be consistent. For pros and cons of the photo grooming, the grooming requirements and format of pricking points photo shall meet C. 2 in Annex C; the grooming requirements and formats of digital photo pricking points shall meet C. 3 in Annex C.

3 The photo control points in the same work area shall be numbered as unified specification, and the number shall not repeated.

7.4 Annotation

7.4.1 The annotation shall meet the following requirements:

1 The annotation may use "field before office" method, "office before field" method, "office and field together" method, etc. For the obvious, scaled ground objects on photo, the qualities and quantities of them could be only explained. The positions and shapes of objects shall be subject to the stereo model of office work. The fair drawing of annotation photo should be implemented with color separation. The interpretation and collection of office work and field work shall be linked effectively.

 2 The annotation shall be implemented according to the requirements of GB/T 7931 or GB/T 13977 and the diagrammatical symbol of GB/T 20257.1 or GB/T 20257.2. The annotation shall include the following items:

  1) Control points.
  2) Settlement places.
  3) Water systems and its outbuildings.
  4) Roads and pipelines.
  5) Transmission lines and communication lines.
  6) Independent ground objects.
  7) Geological exploration points, hydrological and meteorological facilities.
  8) Circumscriptions, classification land boundaries and fences.
  9) The professional irrigational elements such as sea walls, intake sluices, turn-out gates, regulators, lock bolt gate hatches, block tide gates, navigation locks, vegetation, soil texture in dam and reservoir areas of hydro-junctions.

 3 The selection of based drawing scale for annotation of photos or image-maps shall be accurate for interpreting and convenient for fair drawing, and the scale of annotation photos should not be less than 1.5 times the mapping scale.

 4 For the terrain details which need on–the–spot annotation, the positions of the terrain details shall be reached, and the appearance shall be seen, and the conditions shall be asked, and the annotation shall be accurate. The accurate interpretation, clear description, appropriate diagrammatical symbol and correct annotation shall be achieved. The annotation of location features shall be based on the position of photo, and the maximum allowable deviation shall be 0.2mm in the annotation photos or image-maps.

5 After aerial photography, the inexistent ground objects shall be signed with a red "×" on the original image. For the added ground objects after aerial photography, it does not need supplementary survey; if supplementary survey is necessary, it shall be clearly stated in the technical design book.

7.4.2 The annotation work shall meet the following requirements:

1 Before annotation, the scope of annotation line shall be described and edge matching shall be implemented on the photos or images and full map sheet shall be achieved with no holes and no overlap.

2 When the annotation of ground objects is implemented according to the scale, the positions shall be consistent with the images, and the outer contour lines shall be drawn accurately. When annotation of ground objects according to half scale is implemented, the position lines shall be drawn accurately. When the annotation of ground objects not in accordance with the scale is implemented, the image and position points of symbol shall be accurately coincident, and the center of its symbol and the images shall be consistent.

3 For the ground objects and landforms such as embankments, cuttings, scarps, slopes, escarpments terraces, etc. When the length in the map is longer than 10mm, the objects shall be expressed if the relative altitude is greater than 0.5m by adopting 0.5m and 1m contour interval, or the relative altitude is 1m by adopting 2m or above contour interval. When relative altitude is greater than length of a contour interval, the relative altitude should be measured and annotated. When the stereo mapping is used, the annotation may be done in indoor work. However, in the shadow of the valleys and hidden areas, field work is still needed, and the measurement and annotation of relative altitude shall be implemented to 0.1m.

    4 When the annotation is implemented, the water line of rivers, reservoirs, ponds shall be subject to the photo time.

    5 The annotation of rivers, ditches, lakes, ponds, Wells, etc. shall meet the following requirements:

    1) The rivers shall be signed in double lines according to the scale when the width is equal to or greater than 0.5mm in the map; it shall be signed in a single line when the width is less than 0.5mm.

    2) The reservoirs shall be signed in double lines according to scale when the width is equal to or greater than 0.5mm in the map. It shall be signed in 0.5mm single line when the width is less than 0.5mm but greater than 0.3mm. It shall be signed in 0.2mm when the width is less than 0.3mm.

    3) The dry valleys shall be signed in double lines according to scale when the width is equal to or greater than 0.5mm in the map, and it shall be signed in single line when the width is less than 0.5mm. The dry valley depth may not be signed when the width is less than 1m or less than 10mm in the map. When the depth is greater than 2m, it is needed to measure and annotate the ditch depth.

    4) The bank line of lakes and ponds shall be annotated along the edge of the pond. It may not need to be signed when the pond area is less than $4mm^2$. The pond shall be either expressed or ignored while integration of pond is not allowed. But for the large area of pond or the pond segmented by ridge, appropriate integration of pond is allowed.

5) For rivers and canals with fixed flow and the larger channels, flow direction shall be signed. The beginning and end points of navigable rivers shall be mapped. The canal and drainage head above the ground shall be signed.

6) The well outside habitations shall be signed; some of them can be abandoned when there are too many wells. When the well diameter is greater than 2.6mm in the map, its mouth shall be measured and annotated according to its actual shape, and it shall be signed in the symbol of big well according to the scale.

6 The annotation of dikes and gates shall meet the following requirements:

1) The flood bank and sea wall shall be annotated accurately. The dikes higher than 1m shall be signed; however, the dikes with orientation significance below 1m shall also be signed. The dike having significant flood and tidewater control effect, and with the width of dike crown greater than 0.5mm in the map, or the field base greater than 10m, or the dike height greater than 3m shall be signed as stem dike symbol. In addition others shall be annotated as generic dikes. The significant flood banks and sea walls shall be added the name notes.

2) The ground objects of flood bank and sea wall shall be annotated and signed in relevant symbol. The roads shall be

signed by connected double lines, and dikes shall be signedn as terrace. When the roads are signed by the connection single line, there is no sign of road symbols, and the roads shall be denoted the end of the dike.

3) The intake sluices, turn-out gates, regulators, lock bolt gate hatches, block tide locks, etc. shall be annotated. According to construction condition, they shall be signed in symbols of opening to traffic or not opening to traffic respectively, and the divisor whose aperture is greater than 1m shall be signed with this symbol. When the length in the map is greater than 1.9mm, it shall be signed in gate symbol and scaled double lines.

4) The navigation lock shall be annotated, and it shall be signed in symbols of opening to traffic or not opening to traffic respectively according to the traffic situation of the upper gate. When the length in the map is greater than 1.9mm, it shall be signed in gate symbol and scaled double lines. When the distance between two gates is less than 3mm, the major gate is signed only.

7 When the roads signed by double lines go through the dikes, the roads shall be signed as terraces. When the roads signed by single lines go through the general dikes or the main dikes, the roads shall be draw to the end of the dikes.

8 When the ground objects and landform elements have images, the objects and elements shall be drawn as the images accurately. When the ground objects are fuzzy or covered by image or shadow, the intersection method or intercept method adopting obvious ground object point as the start point shall be implemented. For the added ground objects, the relevant distance from the obvious ground objects shall be signed in the an-

notation picture.

9 when the annotation is implemented with the "office before field" method, the clear ground objects and landforms may be mapped in the office work, and the ground objects and landform which are fuzzy or cannot be seen shall be annotated in the field work. The annotation shall mostly focus on the independent ground objects, beacon lights on the bank, wells, water cellars, kilns, pipelines, circumscriptions, roads, water system, vegetation, notes and new residents, buildings, etc., the turning points of power lines and main communication lines shall be annotated accurately. The accuracy, completeness and rationality of ground objects and landforms signed in the original picture shall be checked in annotation. The error shall be removed and corrected one by one, and the incomplete objects shall be completed with supplementary survey, and the unreasonable shall be modified reasonably. The nonexistent ground objects in the original picture shall be signed with a red " × ". All kinds of ground objects which are not shown by notes and video, or are difficult to judge the property shall be annotated supplementarily in the field.

10 When there is no appropriate symbol to express the annotated element, it may be replaced by the similar symbol, or add a new symbol and the instruction.

7.4.3 The annotation and grooming shall meet the following requirements:

1 No matter paper photo or image map or digital image is adopted, the annotation achievement shall be plotted with symbols, words and figures. The grooming shall be clear, and the symbol and annotation shall be clear and easy to read.

2 The annotation and grooming of paper photo or image

map shall use the colorful fair drawing, and the color which means different elements shall be selected according to the stipulation of GB/T 20257.1 or GB/T 20257.2. Black color should be used to describe the ground object and its annotation, brown color should be used to describe the landform and its annotation, green to describe water system, blue to describe double lines river, canals, ditches, lakes and reservoirs, and red to describe the special case instruction. When a simplified schematism is used, in addition to the green still meaning water system, requirements of the simplified schematism symbol shall be followed. Specific requirements can be specified in the design document, and legend explanation shall be used when necessary. The grooming format of the annotation photo can be seen in C.4 of Annex C.

  3 For the annotation of digital image, software of collecting vector data shall be used to plot.

  4 After the field annotation, self-inspection shall be done, and edge matching as required shall be implemented. When edge matching of annotation photo is done, its actual shape and related position shall not be changed. The quality, grade, size, symbol and every annotation of outline, roads, rivers, vegetation, landform, etc. which are outside of edge matching shall be consistent.

7.5 Aerial Image Scanning

7.5.1 The resolution of aerial image scanning shall be determined according to the use of images. If the images are used for multi-purpose, the resolution of the highest requirements shall be chosen, and the following provisions shall be met:

  1 When aerial image scanning is used for digital line graph

mapping and digital elevation model data acquisition, the requirements of elevation measurement accuracy shall be met, and the image resolution shall be estimated according to Equation (7.5.1-1):

$$R_P = 0.8\Delta h b/H \qquad (7.5.1-1)$$

Where:

R_P = image scanning resolution, μm, round-off number;
Δh = tolerance of vertical RMSE of pass points, m;
b = the image baseline length of average heading overlapping, μm;
H = average flying height, m.

2 When aerial image scanning is used for digital orthophoto map drawing, the requirements of orthophoto resolution shall be met, and the image resolution shall be estimated according to Equation (7.5.1-2):

$$R_P = R_O M_O / M_P \qquad (7.5.1-2)$$

Where:

R_P = image scanning resolution, μm;
R_O = orthophoto resolution, μm;
M_O = denominator of mapping scale;
M_P = denominator of photographic scale.

7.5.2 The scanned images shall be clear, and the tone of adjacent photos shall be consistent with each other; the grey level histogram between levels 0 - 255 shall be basically normal distribution; the fiducial marks shall be complete and clear.

7.5.3 The image scanning work shall meet the following requirements:

1 The image scanning work shall be implemented by using the professional image scanner of photogrammetry, and the image scanner shall meet the following requirements:

1) The geometric accuracy is $\pm 2\mu$m.
2) The minimum optical resolution is 7μm.
3) The radiation resolution is not less than 8bit.
4) The optical density is $0.1D - 3.0D$.
5) The dynamic range is not lower than $2.5D$.
6) The size of the minimum output pixel is 7μm.
7) The minimum effective scanning area is 235mm×235mm.

2 For the aerial images of different volumes and time, the pre-scanning work and parameter setting work shall be done well. The parameters of scan brightness, contrast, color saturation and scanning parameters shall be determined during the pre-scanning.

3 According to the parameters of pre-scanning, fragment network scanning of whole volume or the whole section of the photographic film shall be implemented by using of the automatic control device.

4 According to the effect of scanning images, the images shall be enhanced and stretched if necessary.

5 According to the actual situation, the color filter shall be added in the fiducial mark to accomplish the adjustment.

7.6 Digital Aerotriangulation

7.6.1 Before the densification of digital aerotriangulation, the following information shall be collected and analyzed, and data and material shall be confirmed to meet the requirements and the accessibility of data shall be confirmed.

1 The total achievement material of the basic control survey and photo control survey shall be collected.

2 The image data of aerial photographic plate, image or data acquired from digital aerial photography instrument; the aerial pho-

tography schematic figure of the work area, including the division of aerial photography subarea, distribution of routes, and sheet framing; work area image or index image, central point association graph; identification table of aerial photography; auxiliary aerial photographic materials, including data or paper material such as aerial photography site coordinate (GNSS data), image posture parameter (IMU data), etc.

7.6.2 The interior orientation shall meet the following requirements:

1 Fiducial mark coordinate residual absolute value of interior orientation shall not be greater than 0.010mm; the maximum shall not be greater than 0.015mm.

2 Fiducial mark coordinate shall be calculated by using the affine transformation in interior orientation.

3 When the image point is measured, the self-checking adjustment shall be implemented to remove the system error influence of measurement of image point coordinate, such as main image point location, aerial photography instrument distortion, atmospheric refraction, earth curvature, etc.

7.6.3 The relative orientation shall meet the following requirements:

1 The relative orientation precision shall not be greater than the stipulated tolerance in Table 7.6.3, and in especially difficult work area the material or region may be extended to 1.5 times tolerance.

2 The tolerance ΔS、ΔZ of linking difference after model scale submitting to the field shall not be greater than the stipulated tolerance of Equation (7.6.3-1) and Equation (7.6.3-2).

The tolerance of horizontal location linking difference:

$$\Delta S = 0.06 \times M_{image} f \times 10^{-8} \qquad (7.6.3-1)$$

The tolerance of elevation linking difference:

$$\Delta Z = 0.04 \times \frac{M_{image} f}{b} \times 10^{-3} \qquad (7.6.3-2)$$

Where:

M_{image} = denominator of photo scale;
f = aerial photography instrument focal length, mm;
b = image baseline, mm.

Table 7.6.3　Precision requirements of relative orientation

Image type	RMSE of vertical parallax		Residual error of vertical parallax	
	Plains and hills	Mountains and high mountains	Plains and hills	Mountains and high mountains
Image from digitization of aerial image scanning	0.005mm	0.010mm (0.5 times the photo scanning resolution)	0.008mm	0.020mm (1 times the photo scanning resolution)
Image from digital mapping camera	1/3 size of the pixel		2/3 size of the pixel	

3 Every image pair tie point shall be uniformly distributed, and every standard point shall have tie points. When the automatic relative orientation is in process, the number of every image pair tie points shall not be less than 30. The RMSE of vertical parallax of tie point should not be greater than the stipulated tolerance in Table 7.6.3, and in especially difficult work area, the tolerance may be extended to 0.5 times.

4 Choosing objects of pass points or tie points for standard point location shall meet the following requirements, and it shall be clear, obvious, and easy to turn point and measure exactly at the same time.

1) The point location of standard pass point or tie point shall be distributed in the 6 points location of point 1, 2, 3, 4, 5, 6 in Figure 7.6.3. Point 1 and point 2 should be distributed on the ground objects which are 1cm around the principal point of photograph and have clear and obvious image. If the selection of point is difficult, the distance shall not be greater than 1.5cm. The distance from point 3, 4, 5 and 6 to the line across principal point and perpendicular to bearing line shall not be greater than 1cm, and it shall not be greater than 1.5cm in difficult conditions, and the distance from point 3, 4, 5, 6 to bearing line shall be almost equal. For 23cm × 23cm picture format, the distance shall be greater than 4.5cm or 20% of format width. When there is a special requirement of increasing joint strength, the number of tie point shall be added.

2) For the image acquired from digitization of photographic film scanning, the distance from standard pass point or tie point to the edge of image shall not be less than 1cm; for the image acquired from digital mapping camera, on the basis of correcting photogrammetric distortion, the distance from standard pass point or tie point to the edge of image shall not be less than 0.1cm; the distance to each kind of signs shall be greater than 1mm or 50 pixels.

3) The point location of standard pass point or tie point 3, 4, 5 and 6 shall be distributed near the center line of lateral overlap, and mutual position shall form shape of rectangle. When the lateral overlap is too small to ensure the measuring precision, points shall

be selected respectively, and the total vertical distance from two points to overlapping center line shall be longer than 2cm on the image, and tie points in the nonstandard point location shall be added.

4) When the lateral overlap is too large to satisfy the overlap requirements of standard pass point or tie point for adjacent routes, the points shall be selected respectively in 7 – 10cm from the bearing line of 23cm × 23cm picture format or in the position of 30% – 40% of the picture format width, and points shall be turned mutually.

5) In the forest land, the point location of standard pass point or tie point shall be set at the obvious point on open spaces; if it is difficult to choose, it may be set on the treetop, but the images of adjacent routes and stereo image pair shall be clear and easy to confirm its point location.

6) When the points are distributed along river and valley, the relative location among the standard points shall avoid the dangerous circle producing the uncertainty of relative orientation. When the location in plains turns sharply to mountains and high mountains, 1 – 2 terrain feature points shall be added to each image pair on the terrain turning line.

7) The free sheet margin shall have pass point or tie point outside the sheet line.

8) The point locations and numbers of different mapping method, different image scale and different aerial photography linking location shall meet their own requirements, and mutual turning points shall be implemen-

ted.

9) When the standard points are in water, the tie points shall be selected evenly along the bay line.

10) Course tie points should be 3°overlapping, and lateral tie points should be 6°overlapping.

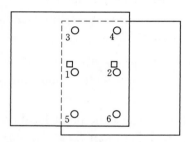

Figure 7.6.3   The standard point distribution of pass point (tie point)

5   When tie points for routes initial link are selected by manual work, the following requirements shall be met:

1) For normal routes, the quantity of tie points between routes shall not be less than 2, and the points should be set in head and tail; for the air strips which have great changes in drift angle, the quantity of rout tie points shall be increased.

2) For long routes, 1 or more tie points shall be added evenly in it, and 1 point should be selected spacing every 10 – 12 image pair.

3) For crossed routes, in the public zone of two routes team, the manual point transfer shall have at least 3 corresponding points, and they shall not be distributed on the same line.

6   When aerotriangulation is implemented with GNSS auxiliary, the tie points between routes may not be selected manually, but it shall pay attention to the usage of auxiliary parameters such as elements of exterior orientation.

7.6.4 Measurement of control points and pass points shall meet the following requirements:

1 The photo control points and check points handed by field work shall be identified and measured. After the implementing comprehensive interpretation, the point positions shall be confirmed accurately according to the puncture, point sketch and explanation.

2 By using the replacement method and repeated adjustment calculation, the gross error of tie points and photo control points shall be detected, and the reliable filed photo control point shall be confirmed, and the gross error points shall be removed.

3 The points with highest precision from tie points which meet the distribution requirements of standard pass points, and the point numbers of pass points shall be arranged automatically or manually according to rules.

4 The confidential check points shall be selected and measured according to needs, and the quantity and numbering rule of confidential check points shall be confirmed in the technical design.

7.6.5 The calculation of absolute orientation and block adjustment shall meet the following requirements:

1 After the calculation of block adjustment, the tolerance of residual shall be 0.75 times the tolerance of RMSE of pass points specified in Table 7.1.2-1 or Table 7.1.2-2; the tolerance of check points error shall be 1.0 times the tolerance of RMSE of pass points; the tolerance of discrepancy of common points in block of flight strips shall be 2.0 times the tolerance of

RMSE of pass points.

2 For the frame perspective aerial photography, the method of bundle block adjustment software shall be adopted in conducting the calculation of aerotriangulation adjustment. For the aerial image by using frame photography and space remote sensing, block adjustment software based on rational polynomial cast (RPC) shall be adopted to calculate the block adjustment.

3 For the GNSS, IMU/GNSS auxiliary aerotriangulation, coordinate of camera station and elements of exterior orientation shall be introduced to implement combined adjustment.

7.6.6 Pass points shall be over-edged in the adjacent region, and the edge matching shall be carried out according to the following requirements:

1 For the edge matching among image, routes, blocks in the same scale and terrain, the horizontal and vertical discrepancy shall be 2.0 times the tolerance of RMSE of pass points specified in Table 7.1.2-1 and Table 7.1.2-2, and the mean value shall be taken as the final used value if the error is less than the tolerance.

2 For the edge matching of the same scale but different terrain, the horizontal discrepancy shall be the sum of horizontal RMSE of pass points of two kinds of terrain specified in Table 7.1.2-1 and Table 7.1.2-2, and the vertical discrepancy shall be the sum of vertical RMSE of pass points of two kinds of terrain specified in Table 7.1.2-1 and Table 7.1.2-2. The horizontal and vertical final used value shall be balanced within tolerance in accordance with the proportion of RMSE.

3 For the edge matching of different scale, the horizontal permissible discrepancy in field shall be the sum of actual length which is obtained from RMSE of pass points of two kinds of scale respectively specified in Table 7.1.2-1 and Table 7.1.2-2; the horizontal final used value shall be balanced within tolerance in accordance with the proportion of RMSE. The vertical permissible discrepancy shall be the sum of RMSE of pass points of two kinds of scale respectively specified in Table 7.1.2-1 and Table 7.1.2-2; the vertical final used value shall be balanced within tolerance in accordance with the proportion of RMSE.

4 For the edge matching of completed or published map, when the discrepancy is less than 1/2 of the tolerance specified in Clauses 1-3 of this article, it shall be subject to the completed or published image. When the range is not less than 1/2 of the tolerance but is less than it, and the mean value shall be taken to be the final used value. When it goes beyond the tolerance, the reason shall be found seriously, and if the completed and published map proved wrong, the current achievement shall be used directly.

5 For the edge matching of horizontal coordinate of common points for different project zone, the conversion of coordinate to the same zone shall be done first, and then, after taking the mean value within tolerance, the mean value shall be converted to coordinate of adjacent projection zone.

6 When the final densified achievement is output, the densified data file shall be output and the consolidation of densified achievement shall be done in accordance with the format required by the next process.

7.6.7 The consolidation of densified achievement shall meet the following requirements:

1 For the densified achievement of point plan sketch in block of flight strips, densified data recorder and automatic aerotriangulation, it shall be groomed and arranged in accordance with unified specification.

 2 When the faults are found in examination and usage of field control points, the relevant results shall be cancelled and the record shall be made, and the name of handler and date shall be recorded at the same time.

 3 When the national control points cannot be pricked exactly and the discrepancy still goes beyond the tolerance after several interpretation and densification of points, the reason shall be found and disposed.

 4 When there is a contradiction between the elevation of river level point and the flow direction, it should be corrected according to the river surface slope ratio, but the corrected value shall not be greater than the vertical RMSE of pass points.

7.7 DLG Data Collection and Production of Map

7.7.1 The orientation modeling shall meet the following requirements:

 1 If densified results of aerotriangulation are available, the results should be chosen to introduce the modeling. If the elements of interior and exterior orientation of image pair are known, elements of interior and exterior orientation should be arranged to accomplish orientation modeling of image pair automatically. The method of the single model orientation shall be selected to accomplish the modeling when the full field control point distribution is implemented.

 2 The absolute value of residual errors for coordinates of interior orientation fiducial mark should not be greater than 0.010mm, and the maximum shall not be greater than 0.015mm.

 3 The relative orientation should use the automatic method, and the number of orientation points shall not be less than

30; manual work shall be adopted in special work, and the points shall be selected in the positions of 6 standard points and the center position of the upside and downside model.

4 The residual vertical parallaxes of relative orientation shall not be greater than the stipulations in Table 7.6.3.

5 The horizontal coordinate errors of absolute orientation in plains and hills should not be greater than 0.2mm in the map while individual shall not be greater than 0.3mm; in mountains and high mountains they should not be greater than 0.3mm on the map while individual shall not be greater than 0.4mm; the height orientation errors and full field control point distribution in plains and hills shall not be greater than 0.6 times the elevation RMSE of elevation point with notes required in Table 7.1.2 - 1 or Table 7.1.2 - 2, and others shall not be greater than 0.75 times the elevation RMSE of pass points required in Table 7.1.2 - 1 or Table 7.1.2 - 2.

6 When the absolute orientation is in process, the photo control points and filed check points handed by field work shall be distinguished and measured, and after comprehensive interpretation of points, the point position shall be confirmed according to the puncture, point position sketch and instruction.

7 When aerotriangulation densified achievements of satellite imagery of pushbroom and remote sensing imagery are introduced to orientation modeling, the relatively large region should be divided into smaller pieces.

8 After modeling, the vertical parallax of single model, the horizontal coordinate and vertical residual error of photo control points and height discrepancy of corresponding image points in adjacent model shall be checked.

9 The work area border of adjacent model shall be seamless spliced.

10 The clipped epipolar line image shall be consistent and matched with the achievements of relative orientation, epipolar resampling and absolute orientation.

7.7.2 The collection of Digital Lined Graph (DLG) data shall meet the following requirements:

1 The RMSE of DLG ground objects in horizontal position, the vertical RMSE of elevation point with notes and the vertical RMSE of sheet contour shall meet the stipulations of Table 7.1.2 − 1 or Table 7.1.2 − 2.

2 The element details and their classification code of DLG shall meet the stipulations of GB/T 13923: the elements in the shape of point, line, surface and annotation shall be classified and collected, and shall be hierarchically saved according to the requirements of professional design.

3 For the stereo collection, the cutting RMSE of contour line and elevation point with notes shall not be greater than $0.04H/b$, and the RMSE of interpretation of ground objects shall not be greater than 0.2mm in the map.

4 For the stereo collection, the surveying and mapping range of image pair shall not exceed 1cm of the orientation point line, and the distance to image edge shall not be less than 1cm, and shall not exceed 3cm of the orientation point line.

5 The point features shall be collected as positioning point, and the second point needs to be collected to confirm its azimuth angle if oriented-point feature exists. The liner element such as road and railway shall be collected as center line, and the liner element signed in scale shall be collected with side boundary. If oriented-line feature exists, it shall be collected as the stipulation that symbol shall be set in right according to the direction of forward motion, and the liner features shall be collected continuously when they meet other features.

6 When the ground objects expressed in accordance with scale are collected, the measuring mark shall pinpoint the contour line exactly; the ground objects expressed in accordance with no scale or half scale, the measuring mark shall pinpoint the positioning point and line exactly.

7 The width of channel and river-way shall be divided into single line or double line in accordance with schema. For the biggish rivers, lakes and reservoirs, the water line should be measured according to the water level when the signs is photographed, and then a water level point every 10 – 15cm in the photo shall be noted.

8 Floodwall and tide barrier shall be measured exactly. For the stem dike, the levee crown elevation shall be measured and noted, and a point every 10 – 15cm shall be noted. The relative altitude of dike shall be measured and noted when it is greater than 1m. The ground objects of floodwall and tide barrier shall be annotated, and the objects shall be expressed in relevant signs.

9 The starting and end points and turning point of line pipe shall be confirmed according to annotation achievement in stereo image, and the line shall be clear.

10 Pattern spot of outline map of vegetation and soil texture shall be acquired according to the image, and the contour line shall be closed.

11 When the contour line is surveyed in field, the measuring mark shall be adopted exactly, and the contour line shall not be drawn omnivorously; in the isoclinal area, the intermediate contour may be interpolated if the interval of index contour is less than 5mm in the map; in isoclinals and change area such as hilltop, piedmont, slop break belt, footstep, etc., half-interval contour shall be surveyed and mapped if the intermediate contour cannot show the geomorphic feature and shape, and the inflection point of contour line shall be in line with the closed and divide line of landform; the half-interval contour shall be drawn if interval of intermediate contour is greater than 5cm in the plains; and the slope line shall be added in valley and concave-convex area.

12 In the visible surface area of forest covered region, it shall be annotated accurately in accordance with surface. If the surface cannot be seen and the contour line is annotated in accordance with tree crown, the correction of tree height shall be taken into consideration.

13 For regions such as intensive forest and shadow of cloud where it is difficult to observe the surface, it shall be expressed in sketch contour when the area in the map is greater than $2cm^2$.

14 When various kinds of physiognomy element signs are interpreted and annotated, the signs shall be in register with corresponding proportion of the relative stereo model exactly. When the signs cannot be registered, the size, shape, and direction shall be altered to make the signs consistent. When the gradient physiognomy symbol is not consistent with the part or total of specification of schema, the part or total shall be changed to adopt the contour line or other appropriate physiognomy symbol shall be used to express.

15 For ground objects and physiognomy with free sheet margin, it shall be measured to 4mm outside the map border.

16 For DLG data among the image pairs, linking and edge

matching shall be done during surveying and mapping. The permissible horizontal and vertical discrepancy of ground object edge matching among image pairs shall be 2.0 times the RMSE tolerance of pass points specified in Table 7.1.2-1 or Table 7.1.2-2. The permissible error of contour line edge matching should be 1 basic contour interval, and it can be extended properly for mountainous region and high mountains, but it shall not be larger than 2 times the vertical RMSE of contour lines. When the basic contour interval is less than 1mm in the map, the edge matching permissible error of contour line shall meet the edge matching permissible error of ground objects.

7.7.3 The edition of DLG data shall meet the following requirements:

1 The data edition shall be in the professional graph edition system. According to the requirements of corresponding scale schema and filed work annotation photo and result, the data shall be edited following the principle of syntheses choice, and the data shall not be distorted or missed while the focal points are outstanding and the gradation is distinct.

2 During the edition of data, the mistakes and omissions of location errors, coverage errors, attribute errors, graph omissions, attribute omissions, annotation omissions shall be eliminated. What's more, the unreasonable phenomenon of mutual contradictory element and uneven lines shall be eliminated.

3 During the data edition with the requirements of creating database, topology mistakes shall be eliminated. The correct topology relation shall be built among the peer elements, and there should not be the hitch point on the line element interface, and there should not be repeated collection or discount for the same line. The area pattern shall close the dimension. When the area pattern intersects with neat lines, the neat lines shall be copied which form a closed polygon.

4 The distributive format of digital diagram product shall meet the requirements of GB/T 17798.

5 The graph express of digital diagram product including sign, line-type, color, annotation, plat grooming, etc. shall be carried out according to the mapping scale required by GB/T 20257.1 or GB/T 20257.2.

6 The output of edited graph file should meet the requirements of GB/T 20257.1 or GB/T 20257.2.

7 The edition of residential area, punctiform ground objects, traffic, line pipe, river system, realm, contour line, vegetation, and noted DLG data shall meet the following requirements:

1) Requirements of residential area edition: 0.2mm interval in the joint of roads and streets shall be kept out. For the constructions built in scarp and slope, they shall be annotated according to their actual locations. When the scarp cannot be annotated exactly, it shall be expressed by shifting, and 0.2mm interval shall be kept out. When the house built above water is coincident with waterside line, the house shall be expressed normally and the waterside line shall be disconnected.

2) Requirements of punctiform ground objects edition: when the distance of two punctiform ground objects is extremely close and it is difficult to annotate both objects, the taller and outstanding one shall be expressed exactly, and the other shall be shifted by 0.2mm, but both shall keep the mutual position relation. When the punctiform ground objects are coinci-

dent with house, roads, river system and others ground objects, other ground objects symbol shall be interrupted and the interval shall be 0.2mm while the integrity of the independent symbol shall be kept.

3) Requirements of traffic edition: when the double lines roads are coincident with the side boundary of construction above the ground such as house and wall, the construction side boundary shall replace the road side boundary, and the interval shall be 0.2mm in the joint of roads side boundary and constructions; when the railway intersects with road horizontally, the railway symbol shall not be interrupted while the road symbol shall be interrupted; when they do not intersect in the same horizontal plane, the relevant bridge symbol shall be noted in the junction of roads; for the embankment (chasm), the side boundary of road and embankment (chasm) shall be mapped, and one shall be shifted by 0.2mm when both are coincident.

4) Requirements of line pipe edition: the power line and communication line of the construction in the city may not be linked, but the direction of ligature shall be mapped; when there are various kinds of lines on the same pylon, the main line shall be mapped, but the direction of lines shall be coherent, and the type of lines shall be clearly demarcated.

5) Requirements of the river system edition: the river shall be interrupted when it comes with a bridge, dam, sluice, etc.; the waterside line may be replaced by scarp line when it is overlapped with a scarp, and the waterside line shall be mapped in the slope foot

when it is overlapped with a slope foot.

6) Requirements of boundaries edition: all the maps noted with country boundary lines shall be implemented as relevant national regulation; there shall not be gap at the corner of boundaries lines, the points, curves or straight lines at the corner shall be drawn; the boundaries shall be annotated as schematism 0.2mm away from the linear ground objects when one side of boundaries are the linear ground objects; when the boundary line is the center of linear ground objects and it cannot be annotated in the center of linear symbols, 3 - 4 symbols every 3 - 5cm along the two sides interlaced shall be drawn, but at the place of boundary cross-points, obvious turning and map-border, the boundary symbols shall not be omitted in order to indicate the trend and position.

7) Requirements of contour lines edition: when the contour lines meet houses and other constructions, such as the double lines roads, embankments, road chasms, holes, scarps, slopes, lakes, double lines rivers, double lines canals, reservoirs, pools, and annotations, shall be interrupted; when the slope direction of contour lines cannot be distinguished, the slope lines shall be added.

8) Requirements of vegetation edition: for the vegetation at the same classification land boundary, its symbol may be distributed equably. When there are greater than two kinds of vegetation at the same classification land boundaries, the symbols may be distributed according to real condition; when classification land

boundary is overlapped with line symbol of material objects on the ground, it may be omitted; when classification land boundary is overlapped with line symbol of no material object on the ground, classification land boundary shall be shifted by 0.2mm and mapped.

9) Requirements of annotation edition: the words annotation shall make the expressed ground objects clear to interpret, and the prefix shall face north; for the names of roads and rivers, they may be spread along with the direction of line curving, and the names shall be placed in side or bottom margin while the names shall be vertical or parallel with the linear objects. The minimum interval among words shall be 0.5mm while the maximum interval should not exceed 8 times the word size. The elevation points with notes should be signed on right of the points, and they shall be 0.5mm far from the point locations. The main ground objects and terrain feature section shall be avoided to be intercepted and covered when the annotation is in process. The annotated prefix of contour level shall point to the peak or the highland and the prefix should not point to the underneath of drawing. For the place with complicated landform, the allocation shall be taken into consideration and the integrality of landform shall be kept. The annotation of map border decoration shall be implemented in accordance with the stipulations of GB/T 20257.1 or GB/T 20257.2.

7.7.4 The sheet edge matching shall meet the following requirements:

  1 For the sheet edge matching of the same scale and same precision, the permissible discrepancy of ground objects horizontal position and contour lines elevation shall be 2 times the relevant RMSE of ground objects horizontal position or elevation RMSE of sheet contour lines in Table 7.1.2 - 1 or Table 7.1.2 - 2 and the mean value in tolerance range shall be adopted to the over-edge.

2 For the sheet edge matching of same scale and different precision, the permissible discrepancy of ground objects horizontal position and contour lines elevation shall be 2 times the sum of relevant RMSE of ground objects horizontal position or the sum of elevation RMSE of sheet contour lines in Table 7.1.2 - 1 or Table 7.1.2 - 2, and balancing and edge matching shall be done according to the proportion of RMSE within tolerance.

3 For the sheet edge matching of different scale, the permissible discrepancy in field of ground objects horizontal position shall be the sum of actual length transformed by RMSE of relevant ground objects horizontal position of the two scale specified in Table 7.1.2 - 1 or Table 7.1.2 - 2. After balancing of actual discrepancy within tolerance according to special ratio of RMSE, the horizontal final used value can be achieved. The permissible elevation discrepancy of contour lines shall be the sum of elevation RMSE of sheet contour lines of two scale specified in Table 7.1.2 - 1 or Table 7.1.2 - 2, and balancing and edge matching shall be done according to special ratio of RMSE within tolerance.

4 For the edge matching of mapped maps or published maps, if the edge matching difference is not greater than the tolerance specified in Clauses 1 - 3 of this article, new maps shall be corrected only; if it is greater than the tolerance, check shall be done seriously, and new maps shall be adopted if it is confirmed that no mistake exists in the new maps.

5 For the edge matching of different scales, the topographic maps of adjacent scales can be over-edged only.

6 When the interval of basic contour lines in the map is less than 1mm, the edge matching tolerance of contour lines may be implemented in accordance with the stipulations of ground objects edge matching tolerance.

7 For the mosaic work of various kinds of ground objects, the indeed shape and relevant positions shall not be changed; the edge matching of landform shall not have any deformation.

7.8 Data Collection and Processing of Digital Elevation Model

7.8.1 The elevations of object DEM grid points shall be close to the surface of an image stereo model, and the maximum shall not exceed twice the RMSE of elevation specified in Table 7.1.3.

7.8.2 For the edge matching of adjacent single model DEM, there shall be at least two grids of overlap area, the elevation permissible discrepancy of DEM corresponding grid points shall be 2 times the RMSE of DEM elevation.

7.8.3 The DEM grid point elevation of static water area shall be coincident, and the DEM grid point elevation of upstream and downstream in the flowing water area shall descend gradiently, and the relation shall be reasonable.

7.8.4 The expanded scope of DEM regular grid points shall get through the limitation of boundary line, and shall meet the following requirements:

  1 The DEM data shall only appear inside the outer boundary line, and the relevant identification symbol of outer boundary shall be 0.

  2 The DEM data shall be interrupted in region consisting of inner boundary line, or be not continuous with the data outside the region. The identification symbol of inner boundary shall be 1. The closed road boundary line, water area boundary line, abrupt change and break line may be used as inner boundary, and may be collected from DLG data.

    3 The boundary line shall be closed polygon, and each DEM data set shall have only one outer boundary line, but may have various inner boundary lines. The different boundary lines may be adjacent, but shall not intersect.

7.8.5 The digital elevation model shall expand by 10mm for all sides on the basis of inner boundary line of standard format, and achievement data shall be created as storage unit by using covered range of rectangle.

7.8.6 The DEM data shall be stored according to the following requirements:

    1 The regular grid point data may be stored by using the following 2 methods.

        1) Store the three-dimensional coordinate (X, Y, Z) of all grid points.

        2) Only store the illustration parameters such as the elevation (Z) of all grid points, horizontal coordinate of left bottom and top right corner of unit (X, Y), size of grid. The elevation storage order of all grid points shall be arranged from west to east and from north to south.

    2 The feature points shall store the three-dimensional coordinate (X, Y, Z) of all points.

    3 The boundary data file may include various boundary line data, and the different boundary lines data may be separated by delimiters. The first line of data file shall include the total number of boundary lines. The first line of each boundary line data shall include the total number of this boundary line points and identification symbols of boundary line. Then the horizontal coordinates (X, Y) of all boundary line points shall be stored in order, and one point shall occupy one line of file while the head and the tail points of the same boundary line shall be coincident.

7.8.7 The edge matching shall be done with adjacent digital elevation model. Fissures shall not appear after edge matching, and the elevation value of overlapping parts shall be consistent; the DEM data of adjacent storage unit shall be linked up smoothly.

7.8.8 The collection work of digital elevation model data shall meet the following requirements:

1 The orientation modeling shall be implemented in accordance with the rules of Article 7.7.1.

2 When the feature points and lines are measured, the observation amplification shall be adopted, and the surveying mark shall cut the surface exactly, and the feature points or lines shall be measured by three-dimensional coordinates. The following factors involved with elevation except the terrain feature lines shall be measured at the same time.

   1) Water line: for static water, water level elevation shall be measured exactly and the water line shall be collected according to the measured elevation, the whole water area shall be built with horizontal triangle according to the elevation, and the DEM grid points shall be assigned according to this elevation. The water line elevation of double lines river shall be assigned sectionally with interpolation method according to the upstream and downstream water level point

elevations.

2) Forest area line: in the forest, when the treetop or tree surface is surveyed in the image space DEM data collection, the ground elevation shall be achieved by using the average height of tree automatically when the object space DEM is created.

3) Non-correlation area line: when the relevant effect of image in some area is no good, the boundary point and inner stake point elevation shall be measured precisely, and the grid point elevation shall be achieved through interpolation method.

3 The acquisition and edition of DEM may adopt the method of object space and image space; by using the method, the DEM generated by stereo matching shall be observed, checked and modified; in the process of edition, the feature points and lines shall be surveyed additionally according to needs to make the object space DEM preferably close to ground; all grid points in scope of contour lines shall be set in the same elevation; height difference shall be substracted from the area with feature.

4 The edge matching of single model shall be done for the production of DEM, and the elevation of grid points of the same name in overlapping area shall be checked. If the elevation discrepancy is larger than 2 times the DEM elevation RMSE, it shall be regarded as overrun. It shall be recorded and the gross error point file shall be formed. The points shall return to the DEM of image pairs respectively, and the whole gross error points shall be shown, and revision surveys of edge matching shall be implemented. The edge matching of the entire single model DEM in the range of sheet shall be completed in order.

5 The mosaic and cropping of DEM sheet shall meet the

following requirements:

1) If the edge matching discrepancy of whole image pairs DEM in sheet scope meet the requirements, the mosaic of sheet DEM shall be implemented; when the mosaic is implemented, the mean value of elevations of all corresponding grid points participating in edge matching shall be regarded as each grid point elevation, and the edge matching precision report of all sides shall be formed.

2) After the mosaic of DEM, the trim shall be implemented quadrangularly according to the start-stop grid point coordinate of the storage unit specified in Article 7.8.6.

6 The examination of DEM data shall meet the following requirements:

1) Examination of object space DEM: the elevation residual of photo control points obtained from stereo model rebuilt shall be checked to find out whether it meets the requirements. On image stereo model, the elevation model of object space DEM grid point shall be checked to find out whether it is close to the ground, and especially whether the gross error point exists.

2) Zero stereo examination by using left/right orthoimage: if the DEM, and DOM data is collected synchronously by using the digital photogrammetric mapping system, the zero stereo effect of left/right orthoimage shall be used to check its DEM quality; if the zero stereo effect image formed by left/right orthoimage has the topographic relief, it means this DEM has error; if the image is vague, check of gross error shall

be done; for the place of gross error and topographic relief parts, marking and rebuilding model shall be done, and the reason shall be found, and at last revision surveys and edit shall be implemented.

3) Check of image pair DEM edge matching: the check shall be implemented to find out whether the all sides edge matching precision of single model DEM is in the range of requirements.

4) Precision check of DEM sheet: by using the elevation check points obtained by field measurement or aerial triangulation, the elevation of relevant points can be obtained by means of DEM interpolation method; after statistics and calculation of the elevation discrepancy, the check shall be implemented to find out whether the RMSE of elevation meet the requirements.

7.9　Production of Digital Orthophoto Map

7.9.1　The image data quality of the digital orthophoto maps produced shall meet the following requirements:

1　Grey wedge and its distribution of DOM images: grey wedge of black-and-white images shall not be less than 8 bit, and grey scale of color images shall not be less than 24 bit; histogram of grayscale shall be presented as normal distribution basically.

2　DOM images shall be clear and easy to read, and the contrast shall be moderate while tone shall be balanced; what's more, obvious trace of image stitching shall not exist.

3　There shall be no double image, unsharpness and fracture texture on DOM images, and the images shall be continuous and intact; the grayscale of images shall not have observable difference. The color of color images shall be balanced and consistent.

4 Ground objects and land features on DOM shall be authentic without such defects as distortion, noise and cloud shadows.

5 Overall appearances of DOM shall be clean, tidy and artistic.

7.9.2 DOM data shall be constitutive of image data, geo-location data and relevant metadata. According to the need, DOM may register geographical name, elevation point with notes and relevant information. And the sheet decoration shall be in process.

7.9.3 Image data as the principal data of DOM shall be stored as TIFF format equipped with geo-location data or Geo-TIFF format.

7.9.4 The geo-location information of DOM data not only contains TITF format or Geo-TIFF format, but also may be described by geo-location data file. When the geo-location information is adopted to locate data file, the file shall contain the following details:

1 Ground simpling distance of image data.

2 Geographic coordinates of south-west corner of image data.

3 Pixel numbers in east-west and north-south direction of image data.

7.9.5 Map borders decoration of DOM shall meet the stipulations in GB/T 20257.1 or GB/T 20257.2. Map borders decoration and text marks on the maps shall be stored in regular vector data format with uniform space coordinates system in accordance with DOM.

7.9.6 DOM data shall use the map-sheet as the storage unit. The data subdivision and designation shall be implemented in accordance with the stipulations in GB/T 20257.1 or GB/T 20257.2. DOM image data should scale out 10mm in scope of inner border's minimum exterior rectangle.

7.9.7 The collection of DOM data may use the method of stereo-model differential rectification or single-scene differential rectification to carry on. The operation of collection of DOM data shall be implemented in accordance with the following requirements:

1 Oriented-modeling shall be implemented in accordance with the stipulations of Article 7.7.1.

2 Parameters of orthophoto maps such as ground sample distance of image and mapping scale shall be set, and method of image resampling shall be chosen. The bi-cubic convolution interpolation method should be adopted.

3 The digital aerial images or epipolar ray images shall be implemented of resampling of differential rectification by using orientation parameters of image elements of interior and exterior orientation and DEM. After modeling, positive and negative should be in the process of ortho-rectification at the same time, and in special situation, the positive and negative may be in the process of ortho-rectification individually.

4 After rectification, the zero-stereo observation and check of positive and negative orthoimage shall be in process, and the obvious topographic relief shall not appear.

5 The implementation of orthoimage mosaic shall meet the following requirements.

    1) All orthoimages which need to be implemented of the mosaic shall be selected according to the map limits.

    2) Between the adjacent images, the mosaic line may be determined manually, or the connection of control points may be adopted as mosaic line to go ahead batch processing automatically. No matter which method is

adopted, the intact images of ground objects which are obtained from mosaic shall be guaranteed; mosaic edge matching difference of adjacent DOM image shall not be greater than two pixels.

6 The image shall be trimmed after it is extended by 10mm in accordance with the scope of inner border minimum exterior rectangle. After the trimming, the raster image file of Geo – Tiff format can be produced, or the sheet orthographic data file and additional information file shall be produced in accordance with the requirements of GB/T 17798.

7 After the rectification, the saturation balance shall be implemented of single-model results. When the saturation balance is in process, a piece of image in working area which contains the colorful information of houses, roads, waters and vegetation shall be chosen. The color information such as color saturation, brightness and contrast shall be adjusted with the image processing software. Then, the qualified image can be achieved as the image sample plot of the whole project. According to the image sample plot, the saturation balance shall be done in lot size by using dodging software.

8 The implementation of DOM data check shall meet the following requirements:

1) Check whether the mathematical foundation of DOM is right and whether the data coverage range meets the requirements.
2) Check whether the edge matching difference of DOM image is in the tolerance.
3) If the left and right images of ortho-rectification are implemented at the same time, the zero-stereo observation and check of positive and negative orthoimage

shall be in process, and the obvious topographic relief shall not appear.

4) Check whether the whole image is clear, tone or color balanced and consistent, and whether obvious image mosaic trace exists.

5) All horizontal check points in scope of DOM shall be measured, and the RMSE of horizontal position shall be counted and precision inspection report shall be acquired to check whether the achievement have met the requirements.

7.10　Airborne LiDAR Scanning Survey

7.10.1 The strips of airborne Light Detection and Ranging (LiDAR) scanning survey shall be designed according to the LiDAR and digital camera technical parameters and the precision of data acquisition, and the following requirements shall be met:

1 Strips lateral overlap of laser data shall be no less than 20%, and the minimum shall not be less than 13%.

2 Strips data files shall include information such as the strip number, the strip order and the system parameters.

3 The scale of the design map shall be chosen in accordance with the stipulations of Table 7.10.1-1.

Table 7.10.1-1　Selection of design map scale

Mapping scale	Design map scale
1:500	1:10000 - 1:25000
1:1000	
1:2000	
1:5000	1:25000 - 1:50000
1:10000	

4 According to the mapping scale and the rules that specified in Table 7.1.8, the ground sample distance (GSD) shall be chosen. Combined aerial camera resolution or pixel size P_i and focal length f with Equation (7.10.1 - 1) and Equation (7.10.1 - 2), the aerial scale denominator M and the average relative flying height H shall be calculated and determined.

$$M = \frac{GSD}{P_i} \times 10000 \quad (7.10.1-1)$$

$$H = Mf/1000 \quad (7.10.1-2)$$

Where:

H = average relative flying height, m;

GSD = image ground sample distance of aerial photography, cm;

P_i = the pixel size of digital camera, μm;

f = the focal distance of digital camera, mm.

5 According to the mapping scale, flying height and the stipulations specified in Table 7.10.1 - 2, the density of laser scanning points shall be chosen, or the density of laser scanning points shall be determined according to the results accuracy requirement.

Table 7.10.1 - 2 **Density of laser scanning points**

Mapping scale	Flying height (m)	Density of laser scanning points (point interval, m)
1:500	500	0.5
1:1000	1000	1.0
1:2000	2000	1.5
1:5000	3000	2.5
1:10000	3900	5.0

Note: The selection of flying height shall meet the formular $H/\cos(\alpha) < L$, where H is maximum relative flying height, α is scanning half angle, and L is the effective range finding of scanner.

6 The accuracy of laser points shall be implemented in accordance with the accuracy of object horizontal positions and the vertical accuracy of elevation points specified in Table 7.1.2－1 and Table 7.1.2－2. In the area of dense vegetation or low reflectivity, the plane and vertical errors of the laser points may be extended by half times.

7.10.2 Ground base stations layout and control connecting survey shall be implemented in accordance with the following requirements:

1 In the implementation of Lidar measurement, 2－3 GNSS base stations shall be set on the ground, and the distance among them should be 30 -50km. The base stations shall synchronously receive GNSS signals in the overall process of flight. GNSS base stations shall meet the requirements of GB/T 18314.

2 In the process of the GNSS data acquisition of ground base station, the elevation mask angle of satellite shall be 15°, and the number of effective observation satellites shall not be less than 4 at the same time interval; and the number of effective observation satellites shall not be less than 9; the number of time interval shall not be less than 2; the interval length shall be no less than 240min; the static sampling interval shall be 30s; the effective observation time of every satellite shallnot be less than 15min in an interval.

3 The ground base stations may conjuntively survey the high grade known points of the nearby known state coordinates.

4 The allowable RMSE of absolute position of ground base station shall be 0.1m, and the allowable height RMSE shall be 0.1m. The allowable side length RMSE shall be 1/500000.

7.10.3 The selection and measurement of calibration field shall be implemented in accordance with the following requirements.

 1 The calibration field should be chosen on the flat hard surface, such as airport runway or a large flat roof house whose size is no less than 20m×80m. Before or after the flight mission, the equipment and instruments shall be checked for flight. Flight mode should be flying face to face respectively vertically and along the runway axis or flat roof house in given height. Every plane every time should be checked for flight, but the face to face route of flight shall be chosen according to the situation of survey area.

 2 The calibration field shall carry out survey in conjunction with GNSS base stations. The permissible RMSE of calibration field control points relative to the horizontal position of the nearby GNSS base stations shall be 0.1mm in the map. The permissible vertical RMSE shall be 1/10 of the basic contour interval.

 3 The calibration field coordination system should be adopted with the current national coordinate system to avoid the system conversion error and enhance equipment calibration accuracy.

7.10.4 The layout and measurement of control points and check points used for mapping coordinate transformation shall be implemented in accordance with the following requirements:

 1 The control points used for mapping coordinate transformation shall be equably distributed around the survey area and the center of the survey area. The highest point and the lowest point of the survey area shall have control points. The conversion control points of the horizontal coordinates transformation points shall not be less than 4, and the number of vertical transformation control points shall be selected according to Table 7.10.4.

Table 7.10.4 **The number of vertical transformation control points**

Mapping scale	The number of transformation points (square area)	The number of conversion points (stripped area)	
1:500	5+S/2	5+L/2	Width≤1km
1:1000	5+S/4	5+L/4	Width≤1km
1:2000	5+S/16	5+L/10	Width≤2km
1:5000	5+S/100	5+L/30	Width≤5km
1:10000	5+S/400	5+L/60	Width≤10km

Note: S is area of survey area, km^2; L is length of survey area, km.

2 The position of the mapping coordinate transformation control points and check points shall meet the following requirements:

    1) The requirements of point positions in GB/T 18314 shall be met.

    2) The original control points which have met the requirements shall be used fully.

    3) The position of the checking points shall be equably distributed in the survey area, and the different height surface and reflection surface shall have checking points.

3 The measurement of the mapping coordinate transformation control points and the checking points shall be implemented according to the following requirements:

    1) The mapping coordinate transformation control points shall carry out survey in conjunction with GNSS base station, and the permissible RMSE relative to the horizontal position of nearby GNSS base station shall be 0.1mm in the map; the permissible vertical RMSE shall be 1/10 of the basic contour interval.

2) The calculation of mapping coordinate transformation control points shall be implemented with the fixed data of geodetic coordinate (B, L, H) of base stations in the current national coordinate, and then the geodetic coordinate (B, L, H) of each control points shall be calculated.

3) The mapping coordinate transformation control points shall have achievements of the mapping coordinate system. If new control points are needed, the mapping coordinate system results shall be calculated besides calculating the current national coordinate system achievements.

4) The horizontal position permissible RMSE of the check points relative to the nearby basic control points shall be 0.1mm in the map; vertical permissible RMSE shall be 1/10 of the basic contour interval. Each map shall have no less than 2 check points, and the total number of check points shall not be less than 50.

7.10.5 The determination of coordinate transformation parameters shall be implemented in accordance with the following requirements:

1 The airborne LiDAR point cloud data shall be converted into the mapping coordinate system and the mapping vertical datum system.

2 When the discrepancy between the check point transformation value and the measured value is not greater than 0.2mm in the map, horizontal coordinate transformation may adopt the Bursa seven-parameter method and Helmet-four-parameter method.

3 When the discrepancy between the check point transformation value and measured value is not greater than 1/5 of the

basic contour interval, the height transformation may use plane fitting method and polynomial surface fitting method.

4 When the survey area has refining achievements of quasi-geoid, the achievements may be directly collected and used.

7.10.6 The airborne laser scanning shall be implemented in accordance with the following requirements:

1 Before the laser scanning survey, the flight control system, laser scanner, digital camera, GNSS antenna and Inertial Measurement Unit (IMU) shall be checked. Only it is confirmed that all the components work normally, can the flight be implemented.

2 When the plane enters the preset air route, the laser radar flight control system controls the infrared laser generator to emit continuum laser to the scan mirror based on the laser equipment operating parameters such as preset scanning mirror oscillating angle, scanning frequency, pulse and so on. By the movement of the plane and the scanning mirror, the laser beams can scan the ground surface and cover the whole area.

3 When the plane enters the default strip, the laser radar flight control system captures image parameters automatically based on the preset digital camera work parameters such as the present camera exposure, shutter speed, ISO value and so on. The camera parameters may be adjusted according to the image quality of the display screen.

7.10.7 The POS data calculation shall be implemented according to the following requirements:

1 According to the airborne GNSS data and ground station GNSS data, the existing national coordinate geodetic coordinate of the positions of airborne GNSS receiving devices shall be determined by adopting DGNSS double differential positioning technology, and then the plane track line combined with IMU data shall be calculated. Every step of the calculated result programs shall have quality inspection report. The laser track data and the camera track data shall be calculated based on the eccentric components of the scanner and the camera.

2 When the survey area is in the area of the built Continuously Operating Reference Station (CORS) of the navigation satellite, it is advisable to use its achievements; the Virtual Reference Station (VRS) may be composed to improve the accuracy of the solution if there are 4 or more base stations on the ground of a survey area and the distribution is appropriate.

7.10.8 The eccentric angle component of the angle of roll (Roll), the angle of pitch (Pitch) and the angle of drift (Heading) shall be calculated in accordance with the flight data of calibration field and the eccentric survey data, and based on this, the ground coordinates of the laser points shall be exported.

7.10.9 Data preprocessing shall meet the following requirements:

1 The POS data of the same flight, the observation data of ground base station, the flight record data, the base station control point data shall be arranged, and then the laser point cloud data which meet the requirements shall be produced.

2 The data of one or more GNSS bases which is the closest to the flight scanning area shall be selected and calculated. The substandard satellite data shall be eliminated. The differential GNSS results and IMU data shall be processed with POS data jointly, and the location of each moment in the flight shall be calculated, and the achievements of the flight track shall be exported.

3 The three-dimensional point cloud data shall be produced and the point cloud data shall be calculated based on the POS data, laser point cloud data and system calibration data. According to the POS data and the relative image data, the image exterior orientation data shall be calculated.

4 The laser data whose system error has been corrected shall be transformed from prevailing national coordinate system to the mapping coordinate system and the mapping height system. The system error among the strips may be balanced by using the checking condition provided by strips lateral overlap zone.

7.10.10 Before producing digital products, the data shall be classified at first: the pure surface data shall be separated while the buildings, vegetation and other non-surface data shall be put into other layer. The pure surface data may generate contour and DEM in the form of TIN or GRID; the digital images may generate DOM after orthographic rectification; the contour and orthophoto may jointly generate the image topographic map; DLG may be generated by the contours which are generated from surface data and the objects which are interpreted from the orthophoto and classified and captured from the laser data.

7.10.11 The accuracy of the digital products shall be checked by using the checking points in the field measurement. If the accuracy does not meet the requirements, the reason shall be analyzed and rectification measures shall be put forward.

7.11 Unmanned Air Vehicle (UAV) Low-altitude Digital Photogrammetry

7.11.1 Unmanned air vehicle low-altitude digital photogrammetry shall choose the mapping scale which can satisfy the precision requirements according to the topography of the survey area, the weather, the flight conditions, the flight platform and the digital camera performance factors.

7.11.2 The flight platform shall be selected according to the following requirements:

 1 The cruising ability of the flight platform shall be longer than the time that meets the need of survey area flying.

 2 The cruising speed shall meet the requirements of the exposure interval when the aerial photography is being implemented. And the speed should not be greater than 120km/h while the fastest should not be greater than 160km/h.

 3 The storage of the autopilot shall satisfy the data storage requirements of the waypoint and exposure.

 4 Navigation and positioning system shall meet the following requirements:

  1) The data output frequency shall not be less than 4Hz.

  2) The double antenna GPS navigation and automatic correction swing angle may be used.

  3) The double-frequency GNSS differential position or precise point position may be used to calculate the actual exposure point coordinates.

 5 When mapping by using the method of IMU/DGNSS data positioning directly, the accuracy of the IMU shall meet the requirements that the roll angle and pitching angle shall not be greater than 0.01°, and the navigation angle shall not be greater than 0.02°.

 6 Other requirements may be executed according to the relevant provisions specified in CH/Z 3002.

7.11.3 The selection and calibration of the digital camera shall meet the following requirements:

1 The digital camera shall meet the following basic requirements:

1) The camera lens shall be prime lens, and focus at infinity with a good stability.
2) It shall be connected firmly between the lens and the camera body, as well as between the camera body and the imaging detector.
3) The imaging detector area array shall not be less than 20 million pixels.
4) The highest shutter speed shall not be less than 1/1000s.
5) The data dynamic range of each image channel shall not be less than 8bit, and a compressed format may be used. The compression ratio shall not be greater than 10 times.
6) The capacity of the camera memory to accommodate the images shall not be less than 500 images.
7) The length of camera battery continuous working shall not be less than 2h.

2 The digital camera calibration shall meet the following requirements:

1) The camera calibration parameters shall include the coordinates of principal points of photograph, principal distance and the distortion equation parameters.
2) The multiple baseline and multi-angle photograph shall be taken towards the calibration field on the ground or in the air when taking camera calibration. Obtain the camera parameter final solution by photogrammetric adjustment method, and statistics accuracy report.

3) Calibration accuracy shall meet the requirement that the permissible RMSE of principal points coordinates shall be $10\mu m$; the permissible RMSE of principal distance shall be $5\mu m$. After fitting by distortion equation and the measured coefficient value, the permissible residual distortion shall be 0.3 pixels.

4) Other calibration requirements shall be implemented according to the requirements specified in CH/T 8021.

7.11.4 The aerial photography plan should adopt a topographic map or image on a scale of 1 : 10000 or a larger scale, or a 3D modeling system may be used. The task range, the accuracy, the usage etc., shall be cleared. The detailed plan shall be developed.

7.11.5 Aerial photography design shall be carried out according to the following requirements:

　　1　The ground sample distance of each photographic subarea datum plane shall be determined according to different scale requirements of aerial photography; the situations of topographic conditions, contour interval, aerial photography base-height ratio, and the usage of image shall be taken into consideration, too. The selection of the ground sample distance shall meet the requirements specified in Table 7.11.5-1.

Table 7.11.5-1　**Selection of the ground simpling distance**

Mapping scale	Value of the ground simpling distance (cm)
1 : 500	$\leqslant 5$
1 : 1000	8 - 10
1 : 2000	15 - 20

　　2　The partition of flight blocks shall meet the following requirements:

1) Boundary of block should be consistent with map border.
2) Elevation difference of landform in flight area shall not be greater than 1/6 of the flight altitude for photography.
3) On the premise of guarantying the linearity of flight course, spacing of blocks should cover the whole aerial photography region completely.
4) If elevation difference of ground varies sharply, difference of terrain feature is remarkable, or there are others special requirements, map border may be broken up to partition the flight blocks.

3 Height of block datum plane shall be determined by topographic relief of blocks, and flight safety conditions. It is suitable to calculate the height in accordance with Equation (7.11.5-1) to Equation (7.11.5-3).

$$h_b = \frac{h_{\bar{h}} + h_{\bar{l}}}{2} \qquad (7.11.5-1)$$

$$h_{\bar{h}} = \sum_{i=1}^{n} h_{ih} \qquad (7.11.5-2)$$

$$h_{\bar{l}} = \sum_{i=1}^{n} h_{il} \qquad (7.11.5-3)$$

Where:

h_{ih}, h_{il} = the height of representative high point and low point in flight block.

4 The layout of flight course should meet the following requirements:

1) The flight course should be completed by parallel flying in east-west direction. In certain conditions, flight may be carried out in north-south direction or along

with directions of lines, rivers, coasts and boundaries.

2) Exposure points should be designed point by point in accordance with topographic relief by using DEM.

3) When the water area and sea area is processed, the condition of principal points of photograph falling in water should be avoided, and all islands covered totally shall be guaranteed and the stereo-pairs shall be built.

5 The selections of season and time of aerial photography shall meet the following requirements:

1) The most favorable meteorological condition in project area shall be chosen, and the negative influence of aerial photography caused by land vegetation and other coverings should be avoided or reduced, and the terrain details can be shown truly by using the image of aerial photography.

2) The time of aerial photography should be determined by solar elevation and multiple of object shadow in the project area of aerial photography in accordance with Table 7. 11. 5 - 2.

Table 7. 11. 5 - 2 **Solar elevation and multiple of object shadow**

Category of topography	Solar elevation (°)	Multiple of object shadow
Plains	>20	<3
Hills and general cities	>25	<2.1
Mountains and large and medium-sized cities	≥40	≤1.2

3) If the flight areas are in desert, gobi, forest, grassland, widespread salt flat and saline-alkali soil, the

aerial photography shall not implemented in 2 hours before and after the local high noon.

4) In the cliffy mountain area and dense high-rise buildings with extraordinary large elevation difference, the time of aerial photography shall be limited in 1h around local high noon, and the aerial photography below clouds may be implemented if conditions allow.

6 For the designed coordinate series of exposure points, the necessary waypoints shall be added in accordance with the requirement of each flight, and the coordinate series shall be laid in the aerial photography system after checking and correcting.

7.11.6 The following implementation requirements of aerial photography shall be met:

1 Taking-off and landing zone and backup site shall be chosen in accordance with the performance requirements of aircraft.

2 Detailed flight plans shall be developed before the implementation of aerial photograph, and some contingency plan shall be worked out as well for possible emergency circumstance.

3 Aerial photography may be implemented under clouds on the premise of ensuring flight safety.

4 Wind power shall not be greater than force 4 when aerial photography is implemented.

5 Other requirements shall be carried out in accordance with the stipulations of CH/Z 3001.

6 If actual exposure point coordinates are needed to calculate by differential GNSS survey, GNSS ground base stations may be laid out nearby.

7.11.7 The following requirements of flying quality shall be met:

1 Degree of longitudinal overlap should be 60% to 80%,

and the smallest shall not less than 53%; degree of lateral overlap should be 15% to 60%, and the smallest individual shall not be less than 8%.

2 Tilt angle of photograph should not be greater than 5° and the largest individual shall not be greater than 12°. The number of pieces whose tilt angle is greater than 8° shall not be greater than 10% of the total quantity. In particularly difficult area, the index should not be greater than 8° and the largest individual shall not be greater than 15°, and the number of pieces whose tilt angle is greater than 10° shall not be greater than 10% of the total quantity.

3 Swing angle of aerial photograph should not be greater than 15°, and the largest individual should not be greater than 30°, if flight course and degree of lateral overlap can meet requirements. In the same flight course, the number of aerial photograph with swing angle greater than 20° shall not be greater than 3 pieces, and the number of aerial photograph with swing angle greater than 15° shall not be greater than 10% of the total quantity. The tilt angle and swing angle of photograph shall not reach the largest value at the same time.

4 The part of longitudinal overlap exceeding working area border side shall not be less than two base lines. The part of lateral overlap exceeding working area border side shall not be less than 50% of picture format. When the surveying of control point of photograph and regular densification in indoor work are not influenced, the part of lateral overlap exceeding working area border side may be extended to no less than 30% of picture format.

5 In the same flight course, difference of flight height of neighboring photograph shall not be greater than 30m, and difference between maximal flight height and minimal flight height shall not be greater than 50m. The difference between actual flight height and designed flight height shall not be greater than 50m.

6 The relative and absolute leaks in aerial photography shall be repaired in time by using the digital camera which is used in the last aerial photography, and the both ends of repairing course shall exceed two photographic baselines which is in the outside of leak.

7 After every flight, the flight logbook of aerial photography shall be filled out in accordance with C. 5 in Annex C.

7.11.8 Image quality shall meet the following requirements:

1 Image shall be clear, rich gradation, contrast moderation, and soft color; the image of tiny surface features matched with the ground sample distance shall be interpreted, and clear stereoscopic model can be built.

2 Defects such as clouds, cloud shadows, smog, widespread reflection of light and stains shall not be existent on image. Though a few defects exist, the connection and mapping of stereoscopic model are not influenced, and the image can be used in mapping line graphic.

3 In the moment of exposure, displacement of image because of aircraft flying shall not be greater than 1 pixel, and the maximum displacement shall not be greater than 1.5 pixels.

4 Mosaic of image shall not have obvious unsharpness, double image and dislocation.

7.11.9 Layout and measurement of control points of photograph shall meet the following requirements:

1 Point location of control point of photograph shall meet the following requirements:

1) Target images of photograph control points shall clear, easy to interpret and stereo measurment. If control point targets are chosen in the places of intersection of tiny linear features (30° to 150°), corner points of outstanding features, punctuate feature center with no greater than 3×3 pixels in raw image, the targets shall be located in the places with less fluctuation of height, relative fixity and easy to exactly locate and measure. Cambered features and shadows shall not be the targets of control points.

2) Locations of vertical control points shall be located in the places with little fluctuation. The intersection and flat hilltop of linear features should be chosen. Narrow gullies, cuspate hilltops, and slopes with obvious fluctuation should not be the targets of vertical control points.

3) The distance from location of photograph control point to photograph edge shall not be less than 150 pixels.

4) When the first and second items are inconsistent with the third item, the first and second items should be implemented.

2 Design and survey of photograph control points shall be implemented in accordance with the requirements in Section 7.2 and Section 7.3.

7.11.10 Annotation of photograph shall be implemented in accordance with the requirements in Section 7.4.

7.11.11 Aerotriangulation shall be implemented in accordance with the following requirements:

1 Relative orientation shall meet the following requirements:

1) RMSE of tie point vertical parallax shall be 2/3 pixels, and the maximum residual error is 4/3 pixels. The tolerance may be extended to 1.5 times in the especially difficult area (large area of desert, gobi, swamp, forest and so on).
2) Discrepancy of bridging of model shall be calculated in accordance with Equation (7.11.11-1) and Equation (7.11.12-2):

$$\Delta S = 0.03 \times M_1 \times 10^{-3} \qquad (7.11.11-1)$$

$$\Delta Z = 0.02 \times \frac{M_1 f}{b} \times 10^{-3} \qquad (7.11.11-2)$$

Where:
ΔS = discrepancy of horizontal location, m;
ΔZ = discrepancy of elevation, m;
M_1 = denominator of a photo scale;
f = focal length of camera, mm;
b = baseline length of photo, mm.

3) Every tie point of image pair shall be laid out equably. When automatic relative orientation is in process, the number of every tie points of image pair should not be less than 30; when manual relative orientation is in process, the number of every tie points of image pair should not be less than 9.
4) On the basis of accurate correction of distortion, the distance from tie points to the edge of images shall not be less than 100 pixels.
5) Other requirements shall be implemented in accordance with the relevant requirements in Section 7.6.

2 Absolute orientations shall meet the following requirements:

1) The self-calibrating block adjustment with additional parameter may be adopted to eliminate system error.
2) Other requirement shall be implemented in accordance with the stipulations in Section 7.6.

7.11.12 Collection and edit of DLG data, DEM data and DOM data shall be implemented in accordance with the stipulations in Section 7.7 to Section 7.9.

7.12 Relevant Data Compilation and Submission

7.12.1 When the aerial and space digital photogrammetry has been completed, the following achievements shall be arranged:

1 The images of aerial and space digital photogrammetry and the relevant technical parameter data.

2 The mapping coordinate system achievements and achievement explanation of photo control points and check points.

3 The original observation data, record handbooks and calculation handbooks of photo control points and check points.

4 The control photos and the description of station flies made by using spreadsheets and digital images.

5 The distribution map of photo control points.

6 The annotated photos or maps by taking analog (papery) or digital images as the carrier.

7 Supplementary survey achievements in field work.

8 The observation data and calculation achievements of aerotriangulation (including achievements such as interior orientation, relative orientation, block of flight strips adjustment and edge matching).

9 The junction map of aerotriangulation block of flight strips, including the details of point locations and numbers of the flight strips, principal points of photograph, photo control points, check points and pass points, etc.

10 The orientation modeling data and handbooks of check record.

11 The data acquisition handbooks, data edition handbooks and inspection record handbooks.

12 Printed output maps.

13 The technical design report, technical summary report and inspection report.

14 Other relevant data.

7.12.2 Data products and relevant documents shall be arranged in accordance with the following contents, and the items shall be registered term by term. Then the achievement list shall be formed. The data shall be handed in after a careful inspection.

1 Data files are shown in Table 7.12.2.

Table 7.12.2 **Data files**

Content	Data format	Storage medium
Data of DLG、DEM、DOM	In accordance with the formats specified in GB/T 17798 or specified commonly used data formats.	Tape/CD/Hard disk
Metadata of DLG、DEM、DOM	In accordance with the formats specified in Chapter 13 or CH/T 1007.	Tape/CD/Hard disk

2 Map files shall include the following contents:

  1) Topographic maps.
  2) Aerial images.
  3) Output maps of DOM achievements.
  4) Output maps of DLG achievements.

3 The documents shall include the following contents:

  1) Achievement lists.

2) Technical design reports and technical summary reports.
3) Quadrangle reports.
4) Inspection reports and receiving reports.

7.12.3 CDs should be the main storage medium of digital achievements while tape, disk or hard disk may also be used. The contents including production tags, production companies, and production times shall be marked on the outer packing.

7.12.4 After the completion of project, the following data shall be handed in:

1 The technical design reports.
2 DLG, DEM, DOM and index plans.
3 The technical summary reports.

8 Terrestrial Laser Scanning and Terrestrial Photogrammetry

8.1 General Requirements

8.1.1 Terrestrial laser scanning and terrestrial photogrammetry may be adopted in mapping a topographic map on a scale of 1:200 – 1:5000, mapping an elevation on a scale of 1:50 – 1:1000, establishing the ground and ground architectural stereo model on a suitable scale, measurement of project quantities and deformation monitoring.

8.1.2 The basic accuracy for ground laser scanning and ground photogrammetry shall meet the stipulations in Table 8.1.2.

Table 8.1.2 Mapping accuracy requirements for ground laser scanning and ground photogrammetry

Error class	Horizontal RMSE (mm in the map)			Height RMSE (m)	
Photo control points	±0.1			$\pm\frac{1}{10}h$	
Feature points	1:200	1:500 – 1:2000	1:5000	—	
	±1.0	±0.8	±0.75	—	
Elevation points with notes	—	—	—	$\pm\frac{1}{2}h$	
Contours	—	—	—	Mountains	High mountains
	—	—	—	$\pm\frac{2}{3}h$	±1h

Notes: 1. h is the basic contour interval, m.
       2. Under difficult conditions the RMSE of height may be extended 0.5 times.

8.1.3 Elevation points with notes shall be selected on outstanding points or relief feature points, and the density shall be determined depending on the capability of the map. 8 – 15 elevation points with notes shall be set every 100cm^2 in the map.

8.1.4 The preparation before work and the instruments calibration work shall be done in accordance with stipulations as follows:

1 The original topographic maps in the survey area and the data of the horizontal and vertical control points within and nearby the surveying area shall be acquired.

2 The requirements of clients, scope of survey area, mapping scale, technology and accuracy requirement shall be realized.

3 The instruments shall be checked before implementation of work.

4 The non-metric camera shall be checked in accordance with the requirements of metric camera. The focal length of the camera, the coordinates of the principal points and the initial value of the distortion difference shall be given.

8.2 Terrestrial Laser Scanning Survey

8.2.1 The layout of scanning base stations shall meet the following requirements:

1 According to the topographical condition, the points of scanning base stations shall be selected, and the scanning view angle of each point should cover all the characteristic points of relief feature points and land feature points.

2 The vision of the scanning base stations shall be wide and the ground shall be stable. It should not be located near any large machinery.

3 If the work area in the field is large, it shall be divided into regions to scan and then to be spliced, and the overlapping degree of different positions and different visual angles of the scanning area should not be less than 10%.

4 Under the premise that the scanning distance is less than the measurement range of equipment, the base station should be laid out on the high positions to make the angle between the scanning ray and ground as large as possible while the ability of scanning would be improved and the scanning gap would be reduced.

8.2.2 The layout and connecting survey of the control points shall be implemented in accordance with the following requirements:

1 The layout of the control points shall be equably distributed on the scanned objects. The number of the scanning control points in small area shall not be less than 4. One control point shall be set at every interval of 400 – 600m in large scanning area. If it is difficult for one side of the scanned object to meet the layout requirements of the control points, or one side of the scanning base station has the condition to lay out control points with higher precision, the control points shall only be set near the scanning base stations. However, the kind of control points shall be easy to identify, and the measurement accuracy shall be improved. The precision shall meet the requirements of mapping control points when the error magnifies near the scanned objects in accordance with the scale of the distance to scanner.

2 Unless the scanned objects have various regular targets with high reflectivity, the survey targets shall be set in the location of the selected control points. The regular spherical targets shall be adopted as the best survey targets. The diameter of the

spheres shall be greater than 3 - 5 times the expected scanning interval. When the survey targets are set in one side of scanning objects, the spheres may be made into hemispherical if the diameters are too large and inconvenient to carry. If the scanners have the function of defining the scanning resolution by area, the smaller diameter survey targets may be used in one side of the scanning objects and the smaller scanning interval is defined specially. The scanning interval should be controlled within 1 - 2cm. The survey targets shall be set up in the open and outstanding positions to ensure no obscured or confusing objects near the targets to improve the space resolution effects in the point cloud.

3 The conjunctive survey of control points shall be implemented in the nearest time of laser scanning, or synchronously.

4 If the scanning base stations have some known coordinates with appropriate precision, the equipment may be set up in the known points and leveled and centered strictly. In the subsequent orientation processing, the known base stations may be used as the control points. If the scanning control point area is defined separately in the same scanning work, the scanning interval shall ensure at least nine laser points available on the survey targets.

8.2.3 The field scanning shall meet the following requirements:

1 The scanning base station shall be far away from the vibration source.

2 The scanners shall be under umbrellas to block the sunshine, so that the over-temperature of the scanners can be avoided. If the temperature is higher than the normal working temperature, measures shall be taken in time to decrease the tem-

perature.

3 The field scanning should be implemented when the surface of scanning objects is dry.

4 Image data shall be captured in the same period of laser scanning. If the external camera is used without fixed installed system, all of the image data may be captured once at the location of the each scanning base station in good weather conditions to reduce the color difference of images.

8.2.4 The mosaic of point cloud data shall meet the following requirements:

1 For a laser scanner which can set the base station coordinates and the visual axis direction of scanner, the scanning achievements of different base stations shall be directly jointed with the unified coordinate systems.

2 For a laser scanner with independent coordinate system as reference of the scanner position and axial direction, various adjacent scanning areas shall be jointed into a big area with at least 10% of the overlapping part, and then the control points are laid out in each large area basis. The layout method of the type of control points can refer to in the Cluase 1 of Article 8.2.2.

3 When the point cloud data is in the process of mosaic, 3 or more common points between two adjacent point cloud data shall be selected, and the relevant parameters shall be calculated automatically in software to join the two pieces of point cloud data together. The parameter of Conv value shall be achieved after the calculation of mosaic. The value shall be less than 1.0×10^{-7} for the high mountains and less than 1.0×10^{-12} for the plains.

8.2.5 Coordinate transformation shall be implemented according to the following requirements:

1 After the mosaic of the point cloud coordinates of the whole piece, the basis of the relative coordinate system shall be in accordance with the datum of the first block of point cloud coordinates. The coordinates shall be transformed if the system shall be incorporated to a certain coordinate system.

2 When the coordinates are transformed, the known control points shall be introduced, and the points shall be associated with the corresponding positions of the point clouds, and the transformation of the whole point cloud data into the unified coordinate system shall be done.

3 In the process of coordinate transformation, the residual errors of the inspection points at the flat shallow hill area should not be greater than 0.2mm in the map, and the maximum shall not be greater than 0.3mm on the map. At mountains and high mountains, the value should not be greater than 0.3mm on the map while the maximum shall not be greater than 0.4mm on the map; the elevations shall not be greater than 0.3 times the basic contour interval.

8.2.6 During the laser scanning, some areas may be blocked by the objects in front of base station so that some laser point cloud data in the rear area will be missed, and in that case another visible location shall be selected in the field as additional scanning station and the supplementary survey shall be implemented. In some special circumstances, the no-prism total station may be used to capture the intensive detail points of loophole area as supplement of the point cloud data.

8.2.7 Annotation of the scanning area shall be implemented according to the following requirements:

1 The annotation should be implemented according to the method of "field before office" specified in Section 7.4.

2 The annotation shall be in units of scanning base station, and the feature information in the relative area shall be noted in the format of sketch.

3 According to the mapping requirements of the corresponding scale, the ground objects, the land feature and their relative attribute information in the scanning area shall be annotated.

4 For the vegetation in the deep valleys and high mountainous area, the vegetation types and heights shall be annotated.

8.2.8 Data processing shall be implemented according to the following requirements:

1 The professional software shall be used to automatically classify and manually edit the point cloud data, and the laser points shall be classified into the ground points and non-ground points.

2 The classified ground points shall be vacuated to generate contour lines. The tolerance of vacuation shall be controlled by 1/3 of the height accuracy of the control point measurements.

3 The ground points shall be imported into 3D modeling software to build 3D grid. Due to the existence of local low reflection area, or area blocked by vegetation or obstacles, the point clouds may have loopholes, and it shall be repaired and edited during the modeling process.

4 For non-topographic survey 3D laser scanning, 3D Dranoff Triangle should be adopted when TIN is built in order to reflect the object space form more accurately.

5 The model decals shall be implemented according to the following steps:

1) If the model data is large, it shall be divided into sev-

eral small parts and the process of model decals shall be implemented separately.

　　2) When small part images are captured, the saturation balance of images shall be done.

　　3) The total model shall be combined and jointed of the small part images after model decals.

　6　During the inspection process of image intrinsic parameters, the permissible error between the calculated focal distance (f) and the real focal distance shall be ±2mm.

8.3　Terrestrial Stereo (Multi-baseline) Photogrammetry

8.3.1　This section shall be adopted in multi-baseline terrestrial stereo photogrammetry. Single-baseline photogrammetry may be implemented in accordance with the requirements specified in GB/T 12979.

8.3.2　The selection of camera station shall meet the following requirements:

　　1　The camera station shall be located in the place with wide vision. The least number of camera stations shall be used to achieve the maximum area.

　　2　The camera station shall be located in the front of the area to be photographed, and parallel or convergent photography shall be adopted in accordance with topographical conditions.

　　3　The tolerance of $Y_{farthest}$ shall meet the calculated value of Equation (8.3.2) when the photographic baselines are chosen.

$$Y_{farthest} = 4Mf \times 20/R_P \qquad (8.3.2)$$

Where:

　　M = denominator of mapping scale;

　　f = focal length of camera, m;

R_P = photograph resolution of camera, μm.

8.3.3 Terrestrial stereo photogrammetry shall be implemented in accordance with the following requirements:

1 When parallel photography is in process, parallel features shall be shot continuously at every camera station, and the front-back image overlapping degree shall be around 80%. When rotational shooting is in process, parallel features shall be divided to several camera stations, and the mode of rotational shooting shall be applied on every camera station and feature shall be covered totally. Image overlapping degree of different camera stations shall be greater than 80%.

2 Images shall be clear with rich gradation, contrast moderation and soft color.

8.3.4 Spot-surveying and the jobs in the following shall be done prior to the layout of camera station:

1 Determining whether parallel photography or convergent photography model is adopted.

2 Determining the scope of surveying area.

3 Looking for control points in field work.

4 Realizing conditions of ground vegetation cover, light, surface slope and transport.

5 Making decision of photographic baseline location in preliminary.

6 Determining the possible photographic loophole.

8.3.5 The jobs of developing technical plan of field work in the following shall be included:

1 Choosing camera station in preliminary and marking it on topographic map in accordance with condition of spot-surveying.

2 Marking scope of photography and stereo mapping on the chosen camera station on topographic map; checking whether overlapping degree meets the requirements or not.

3 Judging whether the longitudinal spacing of photography meets the requirements in accordance with the marked scope of photography and position of camera station on topographic map.

4 Preliminarily selecting control points of photograph and marking them on the map in accordance with scope of stereo mapping and distribution of photograph.

5 Designing connection method of control points of photograph on the map.

6 For the possible photographic leaks, the method of supplementary survey shall be prepared, and the supplementary survey of assistant image pair shall be included.

7 Choosing the best time and route of photography.

8.3.6 Layout and connecting survey of photograph control points shall be implemented in accordance with the following requirements:

1 The layout and connecting survey of control points of photograph shall be implemented in accordance with the requirements in Section 7.2 and Section 7.3.

2 Marks of control points of photograph shall be made in accordance with the requirements of C.6 in Annex C.

3 Control points of photograph shall be measured in accordance with accuracy requirement of mapping control points, and particular points may be extended to the accuracy requirement of observation stations.

8.3.7 The photographic mode shall be chosen in accordance with the existing available photographic equipment while the checking preparatory works of instruments and equipment shall be implemented.

8.3.8 The designed photographic plan shall be strictly carried on in photography, and the enough overlapping degree shall be ensured. The number of camera stations, and photographs as well as data and time shall be recorded on handbook of photography.

8.3.9 Annotation of photograph and supplementary survey of covert area shall be implemented in accordance with the following requirements:

    1 General requirements of photo annotation shall be implemented in accordance with the method of "field before office" specified in Section 7.4.

    2 Photo annotations should be implemented on the camera station of the photo. The ground objects and land features whose contours can be clearly interpreted on photo may not need to do the photo annotation, but the reverse side of photo shall be appended with an annotation text.

    3 The scheme of supplementary survey shall be determined in accordance with the size of covert area. The supplementary survey in field may be implemented if the covert area is small.

8.3.10 Data processing of interior work shall be implemented in accordance with the following requirements:

    1 Preparation of interior work shall include the followings:

        1) Collecting the whole data of field work.

        2) Understanding requirements of tasks, and analyzing data.

        3) Drawing up technical design document or technical design brief of interior work.

        4) Checking instruments of interior work.

        5) Making clear the individual photo control points which need to be densified and marking point positions on photos.

6) Scheduling operational plan of interior work.

2　Modeling and mapping of orientation shall be carried out in accordance with the procedure of relevant multi-baseline mapping software, and the following requirements shall be met:

1) When the relative orientation is in process, at least 6 - 9 orientation points with uniform distribution in different gradations shall be used; residual error and vertical parallax of each points shall not be greater than 0.008mm.

2) Allowable error of horizontal coordinate shall be 0.4mm in the map when the absolute orientation is in process, and error of elevation orientation should not be greater than 1/2 of contour interval.

3) Collection and edit of DLG data, DEM data, and DOM data shall be implemented in accordance with the requirements specified in Sections 7.7 to Section 7.9.

8.4　Relevant Data Compilation and Submission

8.4.1　When terrestrial laser scanning and terrestrial photogrammetry have been completed, the following data shall be arranged:

1　The inspection data of instruments.

2　The record handbooks of control survey.

3　The calculation data and achievement tables of control survey.

4　The laser scanning data and image data.

5　The terrestrial photogrammetry images and annotated photographs.

6 The supplementary survey data for terrestrial photogrammetry loopholes.
7 The achievement tables of densified control points.
8 The indoors orientation handbooks.
9 Technical design documents.
10 The technical summary reports.
11 The data file such as topographic map and DEM.
12 Others.

8.4.2 When terrestrial laser scanning and terrestrial photogrammetry have been completed, the following data shall be submitted:
1 The technical design reports.
2 The achievement tables of control points.
3 The data file such as the topographic map, DEM, etc.
4 The technical summary reports.

9 Remote Sensing Image Interpretation

9.1 General Requirements

9.1.1 This chapter applies to the remote sensing image interpretation of projects in water and hydropower projects construction such as resources and environment investigation, resettlement physical indicator, land utilization investigation, soil erosion monitoring, channel change or shoreline monitoring, and geological disaster monitoring.

9.1.2 The spatial resolution and phase of aerial and satellite images shall adapt to the professional demands. In addition to special needs, aerial images should be color, color infrared or multi-spectral ones while satellite images should be multi-spectral ones.

9.1.3 The accuracy of remote sensing image interpretation may be judged by the correct level of qualitative, positioning and quantitative check methods. And it shall comply with the following stipulations:

  1 The qualitative check method shall judge the land feature type by images. And its interpretation accuracy rate shall be greater than 85%.

  2 The positioning check method shall ensure that the interpreted shapes are consistent with the corresponding image data. It is permitted that there is a shift of 0.3mm in the map between the interpreted shape and its obvious corresponding feature in DOM while a shift of 1.0mm in the map between the interpreted shape and its obscure corresponding feature.

  3 The quantitative check method shall have different con-

cepts and requirements responding to different professions or interpretation objects. Area and the number may be used as the indicators of quantitative checks.

9.1.4 The technical methods for the acquisition, processing, interpretation, mapping and other work of remote sensing information shall be determined by different thematic interpretation contents and accuracy requirements.

9.2　Preparation

9.2.1　Before the remote sensing image interpretation, the technical design for thematic remote sensing image interpretation shall be carried out based on different professional demands. And then, the technical design document shall be prepared.

9.2.2　According to the specific needs of the project, the following information should be selectively collected:

1　Basic geographic data including the latest topographic maps close to the scale of basic interpretation map, DEM, different grades of control points, etc.

2　Aerial and satellite remote sensing data with phase, cloudage, geographical range and other conditions meeting the related requirements.

3　Administrative boundary data and boundaries required by some professionals of water conservancy and hydropower project.

4　Other data in the following:

　　1) Available land investigation and planning data, as well as data about the type, coverage, distribution of vegetation and other data in the project area.

　　2) Regional geological maps, geological disaster maps, etc.

　　3) Data about formation conditions and triggering factors of geological disasters in the project area, including meteor-

ology, hydrology, terrain, formation and structure, earthquake, hydrogeology, engineering geology and human engineering economic activity, etc.

4) Data about status quo and prevention of geological disasters in the project area, including happening time, type, scale, damage condition, investigation, exploration, monitoring, control, rescue, disaster relief and some other work of various geological disasters that happened in history.

5) Data related to humanity, environment, geography, climate and transportation.

6) Other data related to thematic interpretation.

9.3 Remote Sensing Image Processing

9.3.1 Processing of aerial images shall meet the stipulations specified in Chapter 7.

9.3.2 Processing of satellite images shall meet the following requirements:

1 The best band combination shall be chosen according to collected remote sensing data and task demands.

2 In order to facilitate image recognition, different enhancement methods, such as contrast enhancement, edge enhancement, and filtering transformation and so on, shall be used to purposefully enhance the useful information in remote sensing images, highlight thematic feature information and optimize the image quality.

3 Horizontal control may be surveyed by GNSS or other methods, or collected from available DOM and topographic maps whose scale is no less than that of result DOM.

4 DEM of the existing appropriate scale is used for con-

trolling the elevation. The different scale corresponding relationship of DOM and DEM is shown in Table 9.3.2 - 1.

Table 9.3.2 - 1 Scale corresponding relationship of DOM and DEM

DOM scale	DEM scale
1:5000	1:10000
1:10000	1:10000 or 1:50000
1:50000	1:50000

5 Orthorectification shall meet the following requirements:

1) Orthorectification of stereophoto shall comply with the related stipulations specified in Section 7.9.

2) Based on horizontal control points and DEM, the rigorous sensor model or rational function model is used to orthorectify the single scene image. In every scene image the number of control points must not be less than 9 and these points shall be evenly distributed. RMSE of correct control point residuals shall not be larger than the requirements of Table 9.3.2 - 2.

Table 9.3.2 - 2 Requirement for RMSE of correct control point residuals

Type of terrain	Plains and hills (pixel)	Mountains and high mountains (pixel)
RMSE of residuals	1.0	2.0

6 When images are fused, the resolution of multi-spectral images should not exceed 4 times that of panchromatic images. Fused images shall be clear in texture, less distorted in spectral information, and no ghosting and blurring appearance. Image texture of overlap regions shall be consistent. When phase of images is identical or similar, color of them shall be consistent.

7 If a study area involves more than one scene image, all involved images shall be spliced by using seamless mosaic technology of color digital images. For mosaic images whose phase is identical or similar, texture and color of them shall have a natural transition; while for mosaic images whose phase and land features are obviously different, it allows a spectral difference. But the image spectral characteristics in the same land class should be consistent.

9.4 Establishment of Interpretation Signs

9.4.1 Before remote sensing image interpretation, interpretation signs shall be established based on image features.

9.4.2 The establishment of interpretation signs shall meet the following requirements:

1 Through the preliminary interpretation of the interior, the characteristics and distribution of the main deciphers are analyzed in combination with the relevant data, and the characteristics of the interpretation area are grasped, the typical interpretation sample area is established, and the standard of image interpretation is established.

2 Abundant and relatively concentrated area, complex ground objects and difficult terrain interpretation area are selected as the typical interpretation sample areas.

3 Carry out field investigation. before the field investigation, the field investigation plan shall be developed and the suitable investigation route shall be selected.

4 Field investigation, in response to register and record the terrain, the situation of features , the change of land features, characteristic land features, sampling time, sampling place and other things in the whole investigation area, as the basis of establishing regional interpretation signs.

    5 Based on the relationship between target land features and image characteristics, a complete image interpretation of remote sensing image is established through repeated interpretation of images and comparison of the field.

9.5　Remote Sensing Image Interpretation

9.5.1　Remote sensing image interpretation may adopt visual interpretation and computer automatic interpretation. When interpreting remote sensing images, it is necessary to select the appropriate interpretation methods according to different professional requirements.

9.5.2　Visual interpretation shall grasp comprehensive characteristics of target land features and carry out comprehensive analysis based on all direct and indirect interpretation signs in order to improve the quality and accuracy of interpretation. It shall meet the following requirements:

    1 Digital photogrammetry is used to build 3D models while visual interpretation is used to extract the needed information in 3D environment.

    2 GIS software is used to directly interpret and extract information needed from orthorectified aerial or satellite DOM.

9.5.3　Automatic interpretation shall adopt related remote sensing image processing software, and use the supervised classification or unsupervised classification to extract land features based on establishing typical land feature templates.

9.6　Field Verification

9.6.1　Field verification of remote sensing image interpretation shall include the following contents:

    1 Verifying if the type of interpreted land features is correct.

    2 Checking if the boundary of map spots is accurate and correcting the distribution boundary of target land features according to the field verification result.

9.6.2 Before field verification, it is necessary to overlay the interpreted land features on the RS DOM background to make different work base maps for field work whose scale shall be determined by different special demands.

9.6.3 Verification should use the sampling method. And it shall be performed according to the following requirements:

    1 The land features which are difficult to interpret indoors shall be verified emphatically.

    2 According to image characteristics, choose areas with rich spectral information as the verification focus to ensure the diversity and typicality of sampling land feature types.

    3 The verification routes shall be designed based on topographic conditions. It is recommended to carry out verification around settlements and along main roads in flat areas and verification along roads between settlements, or valleys and rivers in mountain and hill.

    4 Verification shall be carried out along the verification route, as well as try to spread to both sides and maximize the verification scope. Every inspection spots shall be reached, seen, asked and drawn.

9.6.4 The verification record of field objects should be detailed and standardized, and the main investigation elements shall not be missed. When the interpretation boundary of target land object is not consistent with the ground, the correct location of the boundary shall be detailedly marked on the working base maps.

9.7 Post Processing of Interpreted Information

9.7.1 According to field verification information and other auxiliary data of interpretation areas, the post processing of extracted thematic information to correct the geographic location, scope and change type of interpreted map spots, shall be carried out.

9.7.2 GIS software shall be used for the topological editing of revised final interpreted data.

9.7.3 Calculation and statistics shall comply with the following requirements:

    1 Based on the specific design requirements of different projects, the area of map spots, length or number of interpreted land features shall be counted.

    2 Unit of calculation and statistics shall be determined by project requirements. Unit of length should adopt meters while that of area adopt hectare or mu.

9.8 Production of Thematic Map

9.8.1 Thematic maps shall include thematic interpretation information, administrative boundaries, geographic names, names of administrative division and other labels, as well as background images. Addition and trade-offs shall be performed based on professional needs.

9.8.2 Mapping scale of thematic maps shall be determined by different professional needs, or area, shape and map size of mapping regions.

9.8.3 Compilation of thematic maps shall comply with the following requirements:

1 Selection and expression of map contents shall be well arranged. Symbols, annotations and other map features shall be correct. The map surface shall be clear and beautiful.

2 For maps without background images, map spots of different land features shall be represented with different colors. For maps with background images, map spots of different land features shall be given different codes.

3 Point and line land features may be symbolized according to the current national map form of different scales. Appropriate symbols for special land features with no symbols in the current national map form may be designed based on professional requirements. And these symbols shall be added to the according legend.

9.8.4 Map subdivision and map-border decoration shall meet the following requirements:

1 Subdivision and numbering of standard framed basic interpretation maps shall comply with the stipulation specified in GB/T 13989.

2 The map-border decoration style of standard framed maps shall comply with the requirements specified in Figure D.0.1 of Annex D.

3 The map-border decoration style of nonstandard framed interpreted maps shall comply with the requirements specified in Figure D.0.2 of Annex D.

9.9 Relevant Data Compilation and Submission

9.9.1 After finishing the remote sensing image interpretation, the following information shall be compiled:

  1 The technical design document.
  2 Image data.
  3 Topographic maps, thematic maps.
  4 Data of calculation and statistics.
  5 Check materials in the interpretation process.
  6 The technical summary report.

9.9.2 After finishing the project, the following information shall be submitted:

  1 The technical design documents.
  2 Thematic maps.
  3 Data of calculation and statistics.
  4 The technical summary reports.

10 Map Compilation

10.1 General Requirement

10.1.1 This chapter suits for compiling topographic maps used by water and hydropower professionals and all kinds of thematic maps.

10.1.2 Design of map compilation shall include the following contents:

1 According to the purpose and use of map compilation, related cartographic data are collected, analyzed and evaluated.

2 Map projection and map scale are determined.

3 Geographical characteristics of mapping regions are researched.

4 Map contents and representation methods are determined.

5 Map symbols are selected and designed.

6 Cartographic generalization principles and map compilation methods are determined.

7 Tests for map compilation are conducted.

8 Technical design documents are prepared.

10.1.3 Map printing shall comply with the stipulation specified in GB/T 14511.

10.2 Compilation of Topographic Map

10.2.1 Topographic maps by field survey shall be compiled based on the requirements specified in Chapter 6 and Chapter 7, while other topographic maps should be compiled by map generalization.

10.2.2 The topographic maps with scales greater than the mapping scale and update information shall be used in compilation of topographic maps. When mapping, some process can be carried out, such as data mosaic, coordinate transformation, content trade-offs and updating, cartographic generalization and editing, etc.

10.2.3 If features in topographic maps greatly change, they shall be updated in real time when mapping.

10.2.4 When compiling topographic maps, in addition to stipulations specified in GB/T 12343, the following rules shall be followed.

1 Coordinates and elevation system of original data shall be converted to the system adopted by the composite map.

2 According to the map sheet range of a mapping scale, the source data shall be transformed and mosaicked. For paper topographic maps, it is necessary to carry out scanning, vectorization, coordinate transformation, mosaic and unifying data format by turns.

3 Composite maps shall adopt the current national map form.

10.3 Compilation of General Geographic Map

10.3.1 The data which are new, rich in contents, close to the mapping scales and need less processing work shall be chosen as the data for the compilation of general geographical maps.

10.3.2 The representation method of general geographic maps shall be appropriate for requirements and purposes of different tasks. The map surface shall be clear and beautiful.

10.3.3 The general geographic map primarily represents various natural elements and human elements, including water system (ocean), landforms, vegetation and land boundaries, settlements, roads, administrative boundaries, etc. When landforms are shown up in hypsometric layer, it is unnecessary to show up the vegetation distribution.

10.3.4 Cartographic generalization shall be carried out based on the usage, scale, mapping areas of the geographic map. Usually, the selected elements should be primary, and the important type features or cartographic targets shall be given priority to reflect.

10.4 Compilation of Thematic Map

10.4.1 The thematic map includes thematic contents and geographical basemaps. In thematic maps, thematic contents shall be emphatically presented. For geographical basemaps, basemap features shall be selectively expressed based on the types of thematic maps, characteristics of mapping regions and map scale.

10.4.2 Topographic maps, general geographical maps, terrain maps or image maps may be the original data to compile the geographical basemaps. Integrate rounding of basemap elements shall be determined by the subject content, scale and characteristics of regions.

10.4.3 When preparing thematic maps, appropriate representation methods and symbols shall be designed according to spatial distributed characteristics, quantity, quality and the dynamic development of thematic elements.

10.4.4 Collection and analysis of cartographic information should be performed in accordance with the following requirements:

    1 The following information about mapping areas should be collected:

1) Map data: maps with smaller scale, topographic maps with large scale, various thematic maps, and national indicative maps.

2) Image data: aerial and satellite images of various scales, etc.

3) Written materials: including various geographical literatures, landform, water resources, climate, soil and various economic zoning data, statistical information from statistical departments and various professional departments, related monographs, etc.

2 For data which is certainly needed for compilation, it's necessary to analyze it, make a concise evaluation, as well as determine the use extent and method of basic material, supplemental material and reference material based on the actual demands. Evaluated contents should include the following contents:

1) Mapping department, foundation of mathematics, mapping time, etc.

2) Political characteristics of maps, accuracy, current characteristics, reliability, completeness of map contents, compliance between geographic contents and the reality, etc.

3) Conformance degree and conversion principles between land features and the symbol, classification, grading in the current national map form.

10.4.5 Before compiling thematic maps, the following contents about mapping areas shall be analyzed and studied:

1 Natural and geographical conditions and socio-economic conditions of mapping areas shall include the following main elements:

1) Terrain, climate, soil, water resources, vegetation, mineral products, etc.

2) Population, gender, age, education level, structure of labor force, traffic conditions, etc.

3) Production situation of industry and agriculture, development imbalance of them, etc.

  2 Objects of map representation are divided into two major categories of features and phenomena. For map features, contents in the ordinary maps, such as classification, distribution, nature of soil, as well as its relationship with other elements and so on, are usually studied. And for research of map phenomena, nature, formation and development laws and some other things shall be understood.

  3 To make the map better suited to the objective situation and the user needs, it's necessary to study the habit standard and recognition ability of professional targets from relevant professional departments as well as traditional representation symbols. The research result is used for the basis of design of thematic map features.

10. 4. 6 The preprocessing of data materials shall include the following contents:

  1 Paper maps shall be scanned and digitized; digital maps shall carry out data format conversion in order to facilitate data editing and processing.

  2 Data processing mainly includes map correction, coordinate projection and transformation, data mosaic, data merging and classification, scale transformation of different map data, etc.

10. 4. 7 Representation of thematic contents shall focus on the map topic. The indicators and representation accuracy shall be selected according to the rules of classification and grading of the thematic contents, and the shapes and representation methods shall be selected according to the map contents and purposes.

10.4.8 Symbols of geographical basemaps should be conventional ones. For symbols of thematic contents, it is necessary to select available symbols or design different graphic symbols based on different professional requirements.

10.5 Atlas Compilation

10.5.1 Atlas shall maintain political ideology, comprehensiveness and integrity of contents, harmonization of interior, and artistry of decoration.

10.5.2 The collected information should be new and accurate.

10.5.3 It shall have a unified standard to unify the map projection, map form and legend, representation method, descriptive text, geographical name index, map sheet size, scale, deadline of referenced information, etc.

10.5.4 For geographical basemaps of atlas, it is suitable to design some kinds of simple, easily comparable scale systems according to the map size and characteristics of different mapping areas. In the same system, it is suitable to adopt one kind of geographical basemap as the basic basemap, and other geographical basemaps are derived from it. With different themes and scales, geographical contents shall make different trade-offs. Basemaps of different systems shall be consistent and coherent in the corresponding contents.

10.6 Relevant Data Compilation and Submission

10.6.1 After map compilation, the following information shall be compiled:

1 The technical design documents.

2 Mapping data.

3 Topographic maps, general geographical maps, thematic maps, data for making atlas and printed results.

4 The technical summary reports.

10.6.2 After completing a project, the following information should be submitted:

1 The technical design documents.

2 Topographic maps, general geographical maps, thematic maps, or printed atlas.

3 The technical summary reports.

11 Special Engineering Survey

11.1 Engineering Survey of Land Acquisition and Resettlement

11.1.1 Engineering survey scope of land acquisition and resettlement shall include reservoir inundated area, construction area and resettlement area.

11.1.2 Engineering survey of resettlement shall include the following contents:

    1 Horizontal and vertical control survey.

    2 Topographic mapping.

    3 Remote sensing interpretation of resettlement physical indicators.

    4 Different kinds of lines, such as boundary of resettlement area, boundary of acquisitive land (for construction), processing lines of urban and town areas and professional projects, etc. shall be set out. In addition, different kinds of boundary monuments and non boundary monuments shall be set out.

    5 The ownership boundary investigation and the present land-use mapping.

    6 Survey of resettlement investigation.

    7 Survey of auxiliary projects in the resettlement area, such as the water supply project, road engineering and power transmission line, etc.

    8 Check the stability of horizontal and vertical control points. Conduct shifting survey or supplementing survey of the control points which will be inundated or have been destroyed.

11.1.3 The mapping scale of topographic maps and present land-use maps may be selected according to Table 11.1.3.

Table 11.1.3 **Mapping scale of topographic maps and present land use maps**

Location	Stage	Scale
Reservoir inundated area	Feasibility study	1 : 2000 – 1 : 10000
	Preliminary design	1 : 1000 – 1 : 2000
Constructions area	Feasibility study	1 : 2000 – 1 : 5000
	Preliminary design	1 : 500 – 1 : 2000
Resettlement area	Preliminary design and resettlement implementation	1 : 500 – 1 : 1000
Cluster production development zone of resettlement area	Preliminary design and resettlement implementation	1 : 500 – 1 : 2000

11.1.4 Before the field survey, the following preparation shall be done:

1 Fully understand the location, survey scope, measured objects, equipments and the starting and ending time of the survey.

2 Determine the expropriation property of inundated line, survey scope, elevation and the terminal position of reservoir back-water, as well as the expropriated land scope of the construction area.

3 Collect the available control survey data and topographic maps of the survey area.

4 Determine the layout plan, class and monument type of horizontal and vertical control network. If the control points in survey area already exits, the technical proposals for checking survey and supplementing survey shall be determined.

5 Contact with the local government within the survey area to discuss a cooperation plan.

11.1.5 The control survey shall meet the following stipulations:

1 In the feasibility study stage, the horizontal and vertical control survey shall be carried out in the reservoir inundated area and the construction area. The existing control data shall be made full use of after being verified, and supplementing survey shall be carried out in the area with insufficient data.

2 Before the survey of inundated, the current basic horizontal and vertical control points in the reservoir shall be shifted and measured above the normal pool level of the reservoir with the same accuracy.

3 In the reservoir inundated area, for the shifting survey or new survey of basic horizontal and vertical control points, not only monuments shall be buried in accordance with the stipulated requirements, but also monuments shall be added in the following locations:

    1) Near the built or proposed relatively large residential area, urban areas, as well as industrial and mining enterprises.

    2) Large farmlands, arable lands, orchards and important cash crop areas.

    3) The convergences of larger tributaries.

    4) Near the hydraulic structures or mineral areas.

    5) Near the historical relics, historical sites, bridges and tunnels.

    6) Landslide monitoring areas and site for proposed hydrological stations.

    7) Levees, bank revetments and scenic spots.

    8) Near the railways, roads and power transmission lines.

4   The class of the head horizontal and vertical control survey shall not be lower than the fourth order.

5   The horizontal control points in reservoir area should be laid out above the inundated line, and the accuracy of height connection survey shall not be lower than the fifth order.

11.1.6   The surveying method, technical requirements and accuracy indicator of the topographic maps required for land acquisition and resettlement project shall comply with the stipulations specified in Chapter 6.

11.1.7   The boundary survey of land acquisition and resettlement area shall meet the following stipulations:

1   All or parts of the following boundaries shall be surveyed according to the project requirements:

   1) Boundary of resettlement area.
   2) Boundary of acquisitive land.
   3) Processing lines of urban areas and special projects.

2   The boundary monuments shall be buried on the field. But when the reservoir inundated area passes areas such as the marshland, puddle, desert, escarpment, permafrost and so on, the boundary monuments may not be buried.

3   The boundary monuments shall be buried along the same elevation line if boundaries of the same type pass through the normal-water segment. And at the backwater curve segment, the elevation of each middle boundary monument shall be calculated by the method of the river-centerline-distance interpolation.

4   The stage-closed boundary monuments shall be set and surveyed at the ends of main streams, tributaries, branch ditches and gullies.

5   The density of reservoir inundation boundary monuments shall be implemented in compliance with the stipulations in

Table 11.1.7-1.

Table 11.1.7-1　Density of boundary monument

NO.	Area passed by inundated line	Density of boundary monument	
		Permanent boundary monument	Temporary boundary monument
1	Large farmlands, arable lands, or forest area with high economic value in plains and hills	Spacing 100 to 200m	Spacing 50m
2	Urban areas, residential areas, industrial and mining enterprises, scenic spots and historical sites	Set 1 monument for each end, set monuments in-between according to the scale and topography	1 monument for every 50m, obvious monuments shall be set on buildings at the main street entrance
3	Small-scaled mountainous cultivated land, sparse houses, forest land, wasteland, grassland	1 monument for 200 – 300m	No less than 2 monuments at each area
4	Collapsing bank, protective area, inundated area, scenic spot	Inter-visible between adjacent monuments, no less than 2 monuments at each area	1 monument for every 50m

Note: Boundary monuments shall be set at the intersection of ownership boundary and flooded line.

6 The boundary monuments shall be buried at every turning point within the range of acquisitive land for project construction; and more monuments shall be added between two adjacent monuments which are invisible to each other or their distance exceeds 100m.

7 The warning boundary monuments for unstable area of reservoir bank shall be set in the surroundings of the unstable area, and the adjacent monuments should be visible to each other.

8 The boundary monuments or signs shall be set on the ground, the base of buildings or at the bottom of trees, where different kinds of boundaries pass.

9 Permanent boundary monuments include four categories: the reinforced concrete monuments, sculpture monuments, steel tube monuments, and signs of water level. Temporary boundary monuments may be made of stake, or be marked on trunks, rocks or walls. Permanent boundary monuments and temporary boundary monuments shall be set according to the requirement specified in Section E. 1of Annex E.

10 In ordinary areas, the reinforced concrete or sculpture boundary monuments should be set, but in sensitive areas like cities, towns and important special projects etc. , signs of water level shall be set additionally.

11 RTK survey, polar coordinate survey with total station, center point survey with leveling instrument or offset survey etc. may be used in the boundary monuments survey.

12 Surveying accuracy of various boundary monuments relative to adjacent control points shall be implemented in compliance with the stipulations specified in Table 11. 1. 7 - 2.

Table 11. 1. 7 - 2   **Surveying accuracy of boundary monuments**

Category of boundary monuments	Area for setting boundary monuments	RMSE of horizontal position	RMSE of height (m)
I	City and town, residential area, industrial and mining enterprises, scenic spots and historical sites, railway, important buildings, road, and the cultivated land where the ground tilt angle is less than 2°	±0. 2mm (in the map)	±0. 1
	Hydro project construction area	±0. 1m	±0. 3

Table 11.1.7-2 (Continued)

Category of boundary monuments	Area for setting boundary monuments	RMSE of horizontal position	RMSE of height (m)
II	The arable land with ground tilt angle of 2° to 6° and the region with great economic value (forest reserve, plantation and livestock farm)	±0.2mm (in the map)	±0.2
III	The arable land with ground tilt angle greater than 6° and the region with certain economic value (general wood, bamboo forest and slope wasteland, etc.)	±0.2mm (in the map)	±0.3

13 The horizontal position and elevation of the boundary monuments shall be surveyed and plotted in the topographic map after they have been embedded. The permissible discrepancy between the measured elevation and the design elevation of the boundary monuments in reservoir inundated area shall be 0.05m. And the permissible discrepancy between horizontal position and design position of land boundary monuments in the construction area shall be 0.05m.

11.1.8 The survey of land ownership boundary investigation shall meet the following stipulations:

1 The following preparations shall be done before the survey of land ownership boundary investigation:

 1) The project develop or ganization shall apply to the competent department of the county government where the surveyed area is located, and then the local government will arrange the land ownership boundary investigation according to the application demands.

2) The working base maps of investigation area shall use the topographic maps, present land use map, or the interpretation result of aerial and satellite images with the same accuracy, which shall meet the design requirements of this stage.

3) Before the survey of land ownership boundary investigation, the local land office at the county or city level, township government or village committee, and the involved organizations of land owners and users shall be informed to participate jointly.

2 The land ownership boundaries which go through obvious land features, landform edges or center lines and can be clearly distinguished in the topographic maps and image maps, may be directly plotted on the topographic maps if the boundaries are unanimously confirmed by all the land ownership parties.

3 As for the land ownership boundaries without any obvious characteristics, they shall be determined through field measurements under the confirmation by all land ownership parties.

4 If there is no objection from either party, the land ownership boundary confirmation letter shall be signed, or the land ownership dispute letter shall be signed by all land ownership parties. If one area is co-owned by several units, it shall be clarified in the Confirmation Letter, in which names, ages, positions and companies of the persons identifying the land ownership boundaries shall be recorded.

5 The boundaries and names of counties, towns and villages shall be marked in each map. State-owned land shall be partitioned into subordinate organizations, and the collective land shall be partitioned into village committees (village groups).

6 The administrative divisions that enclaves belong to shall be marked, and the disputed parts shall be marked with administrative divisions of all disputing parties, for the areas that are not clearly identified, they shall be marked with each administrative division.

11.1.9 Remote sensing interpretation for resettlement physical indicators shall meet the following reqwirements:

1 Interpretation contents include urban areas and rural lands, scattered trees, houses and auxiliary structures, main special facilities, etc.

2 The remote sensing interpretation for resettlement physical indicators should use the method of the stereo interpretation (visual interpretation).

3 In the feasibility study stage, the remote sensing interpretation for resettlement physical indicators shall meet the accuracy of 1 : 2000 topographic maps.

4 The other related requirements of remote sensing interpretation for resettlement physical indicators shall comply with the requirements specified in Chapter 9.

11.1.10 The survey of present land use maps shall meet the following requirements:

1 Digital mapping, aerial and satellite photogrammetry are applicable.

2 Make full use of the available topographic maps of the same scale, or aerial and satellite images with the same accuracy. But the land ownership boundary shall be investigated and surveyed on the spot and the accuracy of topography and land type shall be verified.

3 The subdivision of the present land use maps should adopt the square subdivision of 50cm×50cm. And the maps shall be numbered sequentially from top to bottom and left to right.

As for the areas where topographic maps are available, the original subdivisions and numbers may also be followed.

4 Categories definition and symbol marks of the present land use shall implemented in accordance with the stipulations of GB/T 21010.

5 For the isolated island within the elevation range of reservoir inundated area, the whole area above the survey elevation shall be surveyed.

6 The lowest and highest points of the towns and the residential settlements, streets and their intersections, floor of squares, independent buildings outside the residential settlements in the village, and engineering structures shall be marked with ground elevations.

7 Elevations of main feature points from water conservancy and hydropower project facilities, such as reservoirs or dam ponds, diversion channels, surface water pipelines, pumping stations and hydropower stations shall be surveyed.

8 The ground elevations of various roads, bridges, culverts, ferries or piers, and the pavement of temporary bridge shall be surveyed.

9 For power transmission lines, communication lines, radio and television special lines, etc. quantity of fiber lines or suspension wires and the feature point elevations shall be marked.

10 In the intercropping filed with two or more kinds of crops, the main crop shall be marked and the other crops shall be marked in map spots by percentage.

11 For the villages and residential settlements covering range, the range formed by peripheral boundaries of the village houses may be surveyed and mapped with hachures, and the lowest and highest elevation of the settlements shall be marked as well.

12 The mapping can be omitted when the boundary of land class overlaps with linear land feature or land ownership boundary lines. But the boundary of polygonal land class shall be mapped by shifting 0.2mm.

13 If there are dry farmlands with slope greater than 25° in the map, map spots of the area shall be mapped by boundary of land class, and be marked with the grade of the arable land.

14 If the used basic topographic maps, aerial and satellite images have changes, an on-site revision shall be implemented. The revision technology shall meet the requirements of Chapter 6 as well as the requirements specified in Clause 4 – 13 in this Article.

11.1.11 Shifting survey for surveying control points shall meet the following requirements:

1 The surveying control points shall be implemented with shifting survey according to the principles and requirements of special program reconstruction of resettlement project. And the overall functions of the original control network points in position, accuracy, density, relevance, safety and other aspects shall be totally recovered.

2 The shifting survey points, original control networks or routes as well as the engineering application conditions shall be fully taken into the consideration in the design of technical scheme.

3 The basic control points within the following ranges shall be implemented with shifting survey:

    1) In the land acquisition area and the affected area.

    2) Near the land acquisition area and the affected area in

where the stability of control points will be affected by the construction.

3) The existing surveying control points lose some functions.

4) Supplementary control points are necessary for the shifting survey and rebuilt of control networks or the integrity of routes.

5) Supplementary control points shall be densified to meet the requirements of project construction.

 4 The accuracy and operational requirements for shifting survey of control points shall be implemented in accordance with the stipulations specified in Chapter 4 and Chapter 5.

11.1.12 Survey of auxiliary projects in the resettlement area including the water supply project, road engineering, power transmission line, etc. shall be implemented according to relevant requirements of this code.

11.1.13 The following data shall be sorted out and submitted in the engineering survey of land acquisition and resettlements:

 1 Technical design documents.

 2 Handbooks of field observation.

 3 Calculation documents and outcome tables of horizontal and vertical control survey.

 4 Outcome tables of boundary monuments.

 5 Bailment documents of boundary monuments and land ownership certificate.

 6 Plotting map for boundary monuments.

 7 Layout sketch of horizontal and vertical control network.

 8 The present land use map and map index.

 9 Technical summary report.

11.2 Levee Engineering Survey

11.2.1 Levee engineering survey includes the measurements of various levee projects, such as newly-built, strengthened, expanded and rebuilt ones.

11.2.2 The control survey for the levee engineering survey shall meet the following stipulations:

1 Horizontal and vertical control survey should be added and laid out on the basis of the basic control survey of river basin or the national basic control survey.

2 Control survey shall satisfy the accuracy requirements of topographic mapping, section scale and centerline survey in different design stages. The class of horizontal control survey shall not be lower than the fifth order, and that of vertical control survey shall not be lower than the fourth order.

3 Other technical requirements of horizontal and vertical control survey shall be in compliance with the stipulations specified in Chapter 4 and Chapter 5.

11.2.3 The topographic survey of the levee engineering survey shall meet the following stipulations:

1 In the planning stage, the available topographic maps with scale of 1 : 10000, 1 : 25000, or 1 : 50000 of levee construction area may be used to select routes.

2 The mapping scale and survey range shall satisfy the requirements specified in Table 11.2.3, considering the design stage, engineering properties, topography, landform and other factors.

3 The selection of basic contour interval in topographic maps shall meet the requirements specified in Table 3.0.5-2.

Table 11.2.3 The scale and scope of topographic map survey

Building type	Work stage	Scale	Mapping scope	Remarks
Levee and revetment	Planning	1:10000 – 1:50000	Sideway expanding 100 – 300m like a strip from the center line of levee, and the survey scope should extend outward properly considering the material loading for construction. The longitude extending shall connect to natural highland	The survey scope of sand foundation and back water side for double – foundation shall be widen properly, and should cover and focus on some places like the levee's water-facing side. If it is an eroding beach, the scope shall be extended beyond the thalweg line and encroachment line
	Feasibility study	1:2000 – 1:10000		
	Preliminary design	1:1000 – 1:2000		
	Construction drawing design	1:500 – 1:1000		
Cross building	Planning	1:5000 – 1:10000	Inlet and outlet of buildings, as well as the area connected by both sides	In the preliminary design, lower limit scale should be chosen
	Feasibility study	1:1000 – 1:5000		
	Preliminary design	1:500 – 1:1000		
	Construction drawing design	1:200 – 1:500		

Notes: 1. The topographic map with scale of 1:200 can be surveyed and mapped according to the accuracy requirement of scale of 1:500.
2. It can choose scale of 1:500 for city levee and bank revetment survey.

4 The aerospace photogrammetry and digital topographic survey may be used in topographic survey. The aerospace photogrammetry shall satisfy the corresponding stipulations specified in Chapter 7 while the digital topographic survey shall satisfy the related stipulations specified in Chapter 6. Water area survey shall meet the stipulations specified in Section 11. 8.

11. 2. 4 The center stake survey of centerline shall meet the following requirements:

1 The distance between center stakes of centerline shall not be greater than the stipulated distance specified in Table 11. 2. 4 - 1.

2 The starting stake, ending stake, element stakes of circular curve and deflection point stakes of the levee centerline shall be set and surveyed, and the monuments should be buried if necessary.

3 The center stakes with fixed distance on the centerline of levee shall be set according to the requirements specified in Table 11. 2. 4 - 1. Besides, extra stakes shall be set in the following locations:

Table 11. 2. 4 - 1 **The spacing between center stakes of levee centerline**

Stage	Straight line			Curve
Preliminary design stage and construction drawing design stage	Plains	Hills, mountains	Plains	Hills, mountains
	100	50	50	25
Note: In the feasibility study stage, spacing between center stakes is determined in project task document.				

1) At the crossing of center line and cross-section.

2) At locations with significant topographic changes in the center line.

3) At circular curve stakes.
4) At center points of proposed buildings.
5) At intersections where the center line meets rivers, channels, levees and gullies, respectively.
6) At the location where centerline passes through built sluices, dams, bridges and culverts.
7) At the intersection of the central line and roads.
8) At the residential settlement area, industrial and mining enterprises buildings on the center line and on both sides or within the construction scope.
9) At the boundaries between the open plain regions and mountainous regions or gorges.
10) At both ends of the transition section of designed section changes.

4 The horizontal position accuracy and vertical survey accuracy of center stakes shall satisfy the requirements specified in Table 11.2.4-2 and Table 11.2.4-3 respectively.

Table 11.2.4-2 **Horizontal position accuracy of center stakes**

Unit: m

Levee class	RMSE of center stake position		Discrepancy of stake position	
	Plains, hills	Mountains	Plains, hills	Mountains
First and second class	≤±0.5	≤±1.0	≤±1.0	≤±2.0
Third class and below	≤±1.0	≤±1.5	≤±2.0	≤±3.0

Table 11.2.4-3 **Vertical survey accuracy of center stakes**

Levee class	Closing error	Discrepancy between Two surveys (mm)
First and second class	≤±30	≤50
Third class and below	≤±50	≤100
Note: L is the route length of height survey in km.		

5 RTK survey method, polar coordinate method, method of deflection angles, offset method and liberal method, etc. can be used for horizontal position survey of the levee centerline.

6 If the polar coordinate method and RTK survey method are adopted for measuring the centerline, the following requirements shall be met:

1) The difference between the measured value and the design value of center stake coordinates shall be lower than the position error of the center stake.

2) It is feasible to set out integer stake or supplementary stake without setting the crossing point stakes. It may also only set control stakes on the straight line and curve while the rest stakes surveyed by using the offset method.

3) When the polar coordinate method is adopted, 1 to 2 stake positions that are set out at the previous observation station shall be checked after the station transferred, and the accuracy shall satisfy the stipulations specified in Table 11.2.4-2. When the open traverse method is adopted to survey a few center stakes, the number of open traverse sides shall be no more than 3, and the open traverse shall be closed with the control points with the coordinate closure error less than 10cm.

4) When the single base station RTK method is adopted, the control points used to obtain transformation parameters shall cover the entire setting-out line area and the number of them shall not be less than 4. In addition, the distance between the moving and base stations shall be less than 5km. Another control point

shall be used to check the parameters. And the difference between the observed coordinates and theoretical ones of check point shall be less than 0.7 times the difference of the detected stake positions. Setting-out stake points should not be extrapolated. When network RTK method is used, it shall work within the effective coverage scope.

7 If the method of deflection angles, offset method and liberal station method etc. are adopted to measure the center stake of levee line, the closing error of distance deflection angles shall be less than the specified distance in Table 11.2.4-4.

Table 11.2.4-4 **Closing error of distance deflection angle measurement**

Levee class	Longitude		Lateral (m)		Closing error of angle (″)
	Plains, hills	Mountains	Plains, hills	Mountains	
First and second class	1/2000	1/1000	0.4	0.7	60
Third class and below	1/1000	1/500	0.7	1.0	120

8 If the vertical survey of the levee centerline is implemented, leveling, trigonometric leveling or RTK survey may be adopted, and it shall be started from and ended at the elevation control points of levee lines. Besides the elevation of mark top or stake top, the surface elevation and stake height shall also be measured.

9 Height of the center stakes shall be measured by leveling, which shall meet the requirements specified in Chapter 5. When measuring height of the center stakes by trigonometric leveling, for each distance one round shall measure two readings and vertical angle shall measure one round. For leveling by RTK the requirements are the same with that of the central stake horizontal position survey.

10   When buildings, pipelines, railways, etc. encountered with the survey line, the elevations of them shall be measured according to relative requirements, and the difference between the two measurements shall be less than 2cm.

11.2.5   Section survey shall meet the following requirements:

1   The scale of longitudinal and cross-sections may be selected between 1 : 200 – 1 : 2000.

2   Cross-section surveys shall be conducted on the stakes one by one. In the cross-sectional direction, straight segments shall be perpendicular to the centerline, and circular curve segment shall be perpendicular to its tangential direction.

3   If the centerline of levee intersects with rivers, channels and roads, the intersection angle shall be measured, and the cross-section shall be measured according to the following requirements:

1) If the intersection angle is between 85° to 95°, only one cross-section intersected with rivers or channels may be surveyed along the centerline.

2) If the intersection angle is smaller than 85° or greater than 95°, two cross-sections shall be surveyed. One is perpendicular to the intersected rivers or channels at their center points, and the other is along the centerline.

4   The width of cross-section survey shall satisfy the needs of levee project design, and it should be 100 to 300m extension for each side of centerline.

5 Cross-section survey shall reflect topography and land features; the maximum point distance shall not be greater than 30m in flat areas.

6 Reading accuracy of distance and height difference for cross-sections shall be 0.1m, and its testing comparative difference tolerance shall meet the require ments specified in Table 11.2.5-1.

Table 11.2.5-1 **Comparative difference tolerance of cross-section detection**

Levee class	Distance (m)	Elevation difference (m)
First and second class	$D/100+0.1$	$h/100+D/200+0.1$
Third class and below	$D/50+0.1$	$h/50+D/100+0.1$
Note: D is the horizontal distance between the survey point and the center stake, m; h is the elevation difference between the survey point and the center stake, m.		

7 For the cross-section survey, the RTK survey, polar coordinate survey with total station, and digital terrain model (DTM) may be selected. When the DTM is adopted for collecting the cross-section data, the aerial photography mapping and DTM modeling shall meet the requirements specified in Chapter 7. Besides that, the image control points shall be densified at the location where the vegetation is lush when conducting image control surveying, and survey for the scarp, vegetation, buildings and other topography along the line shall be strengthened at image identification. Cross-section spot check shall be applied to the complex topography area, and the check ratio shall be greater than 5%.

8 Cross-section points for the forward direction of centerline or facing downstream are divided into left and right to modulate section result tables.

9 The mapping scale of vertical and cross-section shall meet the requirements specified in Table 11.2.5-2.

Table 11.2.5-2 **Mapping scales of vertical-section and cross-section**

Map type	Building type	Scale	Remarks
Profile	Levee	Longitude 1:100 - 1:200 Lateral 1:1000 - 1:10000	In the preliminary design stage, lower limit scale should be selected. If the length of levee is larger than 100km, the lateral scale may adopt 1:25000 - 1:50000
Cross section	Levee and bank revetment	Longitude 1:100 - 1:200 Lateral 1:100 - 1:1000	If the cross-section wider than 500m, lateral scale map be 1:2000. The scale of 1:2000 is also applicable for the reinforced old levees

10 Modulate outcome tables of centerline stakes, and produce the vertical section according to the specified map scale after the result is checked no problems.

11 The cross-section mapping shall be conducted in accordance with the following requirements:

1) The mapping scale shall be chosen properly according to the length and height of the cross-section.

2) When multi-cross-sections are needed to be drawn on one map, they shall be ordered by mileage, and arranged from left to right and top to bottom.

3) The centerline stakes of sections in the same column should be on the same vertical line, and the location of centerline stakes shall be marked with eye-catching thick lines, or with "$\triangledown$".

4) When the map is drawn, some space shall be reserved for mapping design section line and marking the number of backfilling and digging at center stakes.

11.2.6 For levee engineering surveying, the following data shall be sorted out and submitted:
1 Technical design document.
2 Control survey data.
3 Levee survey data.
4 The vertical - section, cross - section and plane of levees.
5 Technical summary report.

11.3 Survey for Shoreline Utilization Planning

11.3.1 The main contents of shoreline utilization planning survey include basic control survey, shoreline utilization range and underwater topographic survey, present shoreline utilization investigation and present utilization statistics of different function zones.

11.3.2 Both the horizontal and vertical basic control surveys shall not be lower than the fourth order, and their technical requirements shall be implemented according to the requirements specified in Chapter 4 and Chapter 5.

11.3.3 Considering the importance of planning function and the size of protection scope, topographic map scale should be chosen within 1 : 1000 to 1 : 5000, and topographic map survey shall be implemented according to the requirements specified in Chapter 6 and Chapter 7.

11.3.4 Shoreline feature data shall be collected in fieldwork, and the survey accuracy shall not be lower than the mapping control accuracy.

11.3.5 Shoreline utilization planning survey and investigation shall be implemented according to the following requlrements:

    1 Present levee length survey and present situation investigation shall include the following contents:

        1) Survey for the levee length and mileage, and the current current of reinforced revetment.

        2) Investigation on the organization owning the levee, location of structures crossing levee, levee pavement materials and levee utilization.

        3) If there is no milestone at the starting and ending points of the levee, concrete or large wood stakes shall be laid.

    2 Shoreline length survey and present situation investigation of the current river shall include the following contents:

        1) Survey for the length of outer control line, the riverside control line, and the coordinates of starting and ending points of the shoreline.

        2) Investigation on the organization in charge of shoreline, and the control status (the length and mileage of the vulnerable river section, the bank collapse section, the reinforcement, etc).

    3 Topographic survey and investigation of present shoreline shall satisfy the following requirements:

        1) Reflect the current situation of the control line, the level and tidal flat area, and count the areas of different kinds of land as needed.

        2) Reflect the information such as levee mileage, ownership, etc.

        3) Survey the reclamation scope in accordance with levee

mileage, investigate reclaimed land type and count the area of different kinds of reclamation land.

4) Investigate the population census situation respectively in accordance with levee mileage, administrative region (counties, cities, districts).

4 The survey and investigation of incoming and outgoing tributaries as well as the hydraulic facilities shall include the following contents:

1) Survey the length of river section and river bank occupied by incoming and outgoing tributaries and the characteristic sections of tributaries.

2) Investigate the present situation of the incoming and outgoing tributaries, levee and buildings.

3) Measure the top, and bottom elevation, width of sluice, culvert, dam and other hydraulic structures, and prepare investigation tables as needed.

5 The survey and investigation of non-hydraulic engineering facilities shall include the following contents:

1) As for bridges, docks, harbors, pipelines, cables and other non-water engineering facilities, their location, length of river and shoreline shall be surveged, and area occupied shall be calculated shall be surveyed. If there is a clear scope of protection and management, the range boundaries shall be also surveyed.

2) The height of overhead lines shall be surveyed. Bridges shall be marked with clear height and width. The outlet and inlet of cables shall be surveyed and marked.

3) The technical parameters of facilities shall be collected on site, and investigation tables shall be prepared based on the characteristics of engineering structures.

6 The present utilization statistics of shoreline function zones shall include the following contents:

    1) The shoreline protection area, shoreline reserves, shoreline control utilization area, shoreline development area designed by planning and design unit shall be fitted on the topographic map and be edited to form shoreline function zoning map.

    2) The coordinates of starting and ending points of shoreline control line (river-side control line, outer control line), occupied river length, shoreline length and area of each functional zone in the shoreline function zoning map shall be measured and a statistical table shall be formed.

    3) The present land utilization investigation of each function zone shall be carried out according to the scope of shoreline function zones and the findings within the scope of the shoreline.

11.3.6 For shoreline utilization planning survey, the following data shall be sorted out and submitted:

    1 Technical design document.

    2 Control survey data.

    3 Topographic map and index map, shoreline function zoning map.

    4 Status investigation tables (the levee investigation table of the present river section, the present shoreline investigation table, the present beach investigation table, the tributary investigation table, the present hydraulic and non-hydraulic facility engineering investigation table, the present land use status investigation table of function zones, etc.), the present shoreline utilization table.

5 Technical summary report.

11.4 Water Conveyance Route Survey

11.4.1 Water conveyance route survey shall be conducted in two stages, planning or route selection and design or route location.

11.4.2 Control survey shall meet the following requirements:

1 Horizontal and vertical control should be laid out based on the national basic control network.

2 According to the project scale, control survey may be divided into two levels, the basic control survey and the densified control survey. The class and accuracy of control survey shall meet the requirements specified in Table 11.4.2-1.

Table 11.4.2-1 Basic requirements of control survey

Control level	Horizontal control survey		Vertical control survey	
	The tolerance for the side length relative RMSE of the weakest adjacent points	Survey class	The tolerance of mean accident RMSE of per kilometer height difference	Survey class
Basic control survey	1/40000	Not lower than the fourth order	±3mm	Not lower than the third order
Densified control survey	1/20000	Not lower than the fifth order	±5mm	Not lower than the fourth order

3 For the convenience of restoring the surveyed route and construction setting-out, a certain number of monuments shall be embedded on the route and its vicinity. The monument specification may conform to the related stipulations specified in An-

nex A.

4 The monument points for horizontal and vertical control should be shared, and try to make use of the turning points and kilometer stakes. The distance between monument points may be selected based on Table 11.4.2-2. Large wood stakes shall be buried on the turning points if no monument is buried on the rounte.

Table 11.4.2-2 Spacing of monument points for horizontal and vertical control

Stage		Horizontal control point	Vertical control point
Planning stage		2 monuments every 3-5km	Vertical points for connecting survey of the horizontal control points
Design stage	Route vicinity	3 monuments every 5km	1 monument every 1-3km
	Main buildings	2 monuments at each site	1 monument at each site

5 Horizontal control is recommended to be surveyed by GNSS, but other methods are also applicable. Survey technology shall meet the stipulations specified in Chapter 4.

6 Vertical control should be surveyed by leveling. The fourth order leveling may also be surveyed by trigonometric leveling of electromagnetic distance measurement. Survey technology shall meet the stipulations specified in Chapter 5.

11.4.3 Topographic survey shall meet the following stipulations:

1 In the planning stage the strip topographic map at 1∶2000-1∶10000 scale shall be surveyed, while in the design stage strip topographic maps at scale of 1∶1000 or 1∶2000 and topographic maps at scale of 1 : 500 or 1 : 1000 for building area shall be surveyed.

    2 The contour interval of topographic maps shall meet the stipulations specified in Table 3.0.5 - 2.

    3 Topographic survey may be conducted by the method of digital topographic survey or aerospace photogrammetry. Digital topographic survey shall meet the relevant stipulations specified in Chapter 6. Aerospace photogrammetry shall meet the relevant stipulations specified in Chapter 7.

11.4.4 Route selection and location survey shall meet the following stipulations:

    1 If the route in field work is selected, the route turning points, center location of sluice, culvert and other route buildings shall be marked by wood stakes fixed on site.

    2 In the design stage, for layout and survey of the circular curves their starting, middle and ending points shall be measured and vertical control points shall be set properly outside of the construction area.

    3 The survey accuracy for center traverse points, center line stakes and cross - section points of water conveyance route shall meet the stipulations specified in Table 11.4.4.

Table 11.4.4 **The survey accuracy of center traverse points, centerline stakes and cross - section points**　　Unit: m

Point type	Center traverse points and centerline stakes		Cross - section points	
	Plains, hills	Mountains, high mountains	Plains, hills	Mountains, high mountains
RMSE of point location to adjacent basic horizontal control points	2.0	2.0	—	

Table 11.4.4 (Continued)   Unit: m

Point type	Center traverse points and centerline stakes		Cross section points	
	Plains, hills	Mountains, high mountains	Plains, hills	Mountains, high mountains
RMSE of horizontal position to centerline stakes	—	—	1.5	2.0
RMSE of height to adjacent basic height control points	0.1		0.3	

Note: The height survey accuracy may be relaxed 0.5 times if only the requirements in the planning stage are considered.

4 Route selection and location should be surveyed by the methods of RTK survey or center traverse survey.

11.4.5 Center traverse survey shall be implemented in accordance with the following stipulations:

1 The location RMSE and height RMSE of center traverse points shall be in compliance with the stipulations specified in Table 11.4.4. The allowable point location RMSE of center traverse's starting and closing points relative to adjacent basic or densification horizontal control points is 1m, and that to adjacent basic or vertical control pass points is 0.05m.

2 The horizontal position, elevation as well as the vertical-section mileage of the center traverse points may be surveyed together. The length of annexed center traverse between two high class points and elevation route shall not exceed the stipulations specified in Table 11.4.5.

3 For center traverse points, besides the elevation of mark top or stake top, the surface elevation or stake height shall also be surveyed.

Table 11.4.5 The length of center traverse and height route   Unit: km

Connecting center traverse	Height route	
	Fourth class	Fifth class
50 (or 50 turning angles)	80	30

4 On the water conveyance route, besides the fifty-meter stake, hectometer stake and kilometer stone, additional stakes shall also be set at the locations specified in Clause 3 of Article 11.2.4. All additional stakes shall be numbered by kilometer. The mileage, stake top height and surface elevation of each point shall be surveyed.

5 Points on the circular curve may be surveyed by methods of RTK, deflection angle, long chord angle, tangent offset, polar coordinate and chord deflection distance. The circular curve survey shall meet the following requirements:

  1) Comparing the route distance surveyed along the circular curve stake with the computed theoretical distance, the difference shall not be greater than 1/1000 of the curve length.

  2) The lateral permissible error of curve survey is 0.2m.

11.4.6 The vertical-section and cross-section survey shall meet the following stipulations:

  1 The spacing between vertical-section points and cross-section of the water conveyance route shall be determined according to Table 11.4.6-1.

  2 The spacing between cross-sections shall satisfy the stipulations specified in Table 11.4.6-1. In addition, the cross-section shall also be surveyed on the additional stake location specified in Clause 4 of Article 11.4.5.

Table 11.4.6-1 Spacing between veritical-section and cross-section of water conveyance route   Unit: m

Stage	Cross-section spacing		Distance of vertical-section points	
	Plains	Hills, mountains	Plains	Hills, mountains
Planning	200-1000	100-500	The spacing between basic points shall be the same as that of cross-sections, and extra points shall be added in special positions	
Design	100-200	50-100		

3 If the centerline of water conveyance route intersects with rivers, channels or roads, the survey shall be in compliance with the requirements specified in Clause 3 of Article 11.2.5.

4 The density of cross-section points shall be big enough to fully reflect the topographic change. The maximum point distance shall not be greater than 30m in flat areas. Every topographic change point shall be measured in field.

5 The spacing of cross-section points is calculated from the centerline stake, and the section is divided into left and right parts in the direction of the centerline, the cross-section survey table is then modified.

6 The mapping scale of vertical-section and cross section may be selected based on the requirements set out in Table 11.4.6-2 and Table 11.4.6-3.

Table 11.4.6-2 Mapping scale of vertical-section

Stage	Mapping scale of vertical-section		
	Horizontal scale	Vertical scale	
		Plains	Hills, mountains
Planning	1:10000-1:50000	1:50-1:200	1:100-1:500
Design	1:5000-1:25000		

Table 11.4.6-3　Mapping scale of cross-section

Length of cross-section	Mapping scale of cross-section		
	Horizontal scale	Vertical scale	
		Plains	Hills, mountains
<100	>1:500	1:50 – 1:100	1:100 – 1:200
100 – 200	1:500 – 1:000		
200 – 500	1:1000 – 1:2000	1:50 – 1:200	1:100-1:500
>500	<1:2000		

7 Mapping of vertical-section and cross-section shall meet the stipulations specified in Clause 10 and Clause 11 of Article 11.2.5.

8 The route plane map may be made using existing topographic maps as base maps. The starting point, ending point and turning points of the route, the starting point and ending point of the curve, kilometer stone, and supplementary land features surveyed along the route shall be mapped on the base map one by one. The turning angle I, curve radius R, curve length L and tangent length T of the circular curve shall be marked near the turning point.

11.4.7 Land acquistion and resettlement survey shall be in compliance with the stipulations of Section 11.1.

11.4.8 For water conveyance route survey, the following data shall be sorted out and submitted:

1　Technical design documents.
2　Control survey data.
3　Route survey data.
4　Vertical-section, cross-section and plain map of

route.

5 Technical summary reports.

11.5 Survey of Electric Power Transmission Line

11.5.1 This section applies to the electric power transmission line survey between the power grid transformer station or power station and the construction area during the construction stage of water conservancy and hydropower project. The main work consists of route selection and alignment; including route control survey, strip topographic mapping, section survey, pole (tower) survey, and route crossover survey, etc.

11.5.2 The electric power transmission line survey shall meet the following general stipulations:

1 The horizontal coordinate and elevation system of route control survey shall be selected according to stipulations specified in Article 3.0.1 – Article 3.0.4.

2 The mapping scale of route may be selected according to the following requirements:

1) The topographic map scale of construction site shall be 1 : 200 or 1 : 500. The 1 : 200 topographic map may be surveyed according to the accuracy requirements of 1 : 500 topographic map.

2) If the overhead transmission route passes through the agreement plot or planning plot of urban area, the strip topographic map of scale 1 : 1000 – 1 : 2000 shall be surveyed according to the requirements of the local planning department.

3) If the cross – section shall be surveyed for the overhead transmission route, the scale of 1 : 200 or 1 : 500

should be selected.

3 If the route crosses existing roads, pipelines or power transmission routes, the cross angle, the horizontal position and elevation of cross point as well as clear height or negative height of cross point shall be surveyed as required.

4 As for the land features on the X-axis of vertical – section, the location, elevation and necessary height shall be surveyed in field as required.

5 Connecting survey for the coordinates and elevation of the route shall be in compliance with the following requirements:

1) If the route is close to or pass through the planning area, industrial and mining areas, military facilities district, signal transmission and reception station or preservation area of cultural relics, etc. connecting survey of the plane coordinate shall be conducted if necessary, with the accuracy of 0.6mm in the map.

2) If the route goes through rivers, lakes, reservoirs, drainage networks, flooded areas, or through railways, highways, overhead pipes and other buildings under planning or construction, connection survey of elevation shall be conducted according to the mapping control class if necessary.

6 Fixed stakes shall be buried in starting, ending and turning points of the route.

7 Before construction, the confirmed route shall be inspected. Only if the route accuracy satisfies the requirements, may the work of setting out and construction be carried out.

11.5.3 The selection of electric power transmission route shall meet the following stipulations:

1 Collect the data of 1 : 10000 – 1 : 50000 topographic

maps and aerial photographs, satellite images along the route, horizontal and vertical control points, as well as large-scale topographic maps of planning area, crowded section, complex topographic area, large crossover area and other special area for indoor selection of route.

2 Assisting designers compare the indoor route selection schemes through field investigation. The areas affecting the route selection, including agreement areas, crowded areas, large crossovers, important crossovers, and areas with complex topography, geology, hydrology, and weather condition are the key places for field investigation, and shall be surveyed by instruments if necessary.

3 According to the approved route scheme, the selection shall be confirmed cooperating with the design contents. If the route goes through the urban area, agreement area and area with heavy land features, or if the adopted coordinate and elevation system are inconsistent with the local ones, connecting survey of coordinate and elevation system shall be conducted to obtain the conversion relationship, and the related land features and landform shall be surveyed.

11.5.4 Control survey shall meet the following stipulations:

1 The accuracy indexes of horizontal and vertical control points should satisfy that of the mapping control points, and may also be determined by the design requirements.

2 Horizontal control survey of the route should use the GNSS method or traverse survey, and the control points should be set close to the route. Vertical control survey should use the methods of leveling or the EDM trigonometric leveling, and the control points should be set close to the route.

3 The monument may be embedded in the location of horizontal and vertical control points as needed.

11.5.5 Alignment survey shall be implemented in accordance with the following stipulations:

1 Methods of direct and indirect route alignments may be used for route alignment. The RTK method and mid-division with telescope in normal and reversed position method may be used for direct route alignment survey while the rectangle method and isosceles-triangle method may be used for indirect route alignment. The direction deviation from the straight line shall be within $180°\pm1'$.

2 The main technical requirements of route alignment survey shall meet the stipulations specified in Table 11.5.5.

Table 11.5.5 **Main technical requirements of route alignment survey**

Alignment method	Device accuracy class	Alignment error of instrument	Horizontal offsetting of bubbles	Point location difference by facing left and right	Distance relative error
Direct route alignment	6″	≤3mm	≤1grid	Every 100m, ≤60mm	1/5000
Indirect route alignment	6″	≤3mm	≤1grid	Every 10m, ≤3mm	1/2000

Note: If RTK method is used, the accuracy shall not be lower than that of the survey station point.

3 The horizontal angle of corner stake may be measured by device of grade 2″ or 6″ with one round, and the tolerance of 2C mutual deviation is 60″.

4 The permissible relative error of distance measurement between alignment stakes shall be 1/200 in the same direction, and 1/150 in reciprocal observation. The larger span should be

taken care by electromagnetic distance measurement, with the permissible relative error of distance measurement 1/span (m).

  5 The permissible elevation difference in reciprocal observation between route alignment stakes is 0.1S. S is the distance between route alignment stakes with the unit as 100m. The elevation difference of route over larger span should be surveyed by the electromagnetic distance measurement with mapping trigonometric leveling.

11.5.6 Plane and vertical – section survey shall meet the following stipulations:

  1 Collect or measure the relative position plane of route starting, ending point and transformer substation according to the design requirements.

  2 The plane positions of affected buildings (structures), roads, pipelines, rivers, reservoirs and ponds, channels, terraced fields and underground cables within 50m on each side of the route center line shall be surveyed on-site.

  3 If the route goes through the forest, orchards, nurseries, areas of crops and economic crops, the boundaries shall be surveyed in field, and the names, species and heights of the crops shall be marked as well.

  4 The method of RTK or polar coordinate, and electromagnetic distance measurement may be adopted in vertical – section survey.

  5 The interval of vertical – section points should be no greater than 50m. Extra points shall be added appropriately at the topography variation position. At least three vertical – section points shall be selected on individual hills.

  6 In the areas where the transmission line may danger the ground, the points on vertical – section should be properly densi-

fied.

7 If the power transmission route passes through valleys, deep gullies, where the line height to ground is safe, the vertical – section line may be interrupted.

8 If the arrangement of power transmission lines is wide and the side-line ground is 0.5m higher than the measured centerline ground, the side – line vertical – section shall be surveyed.

9 The land feature on the X-axis of vertical – section shall be surveyed in compliance with the stipulations specified in Clause 4 of Article 11.5.2. In addition, the vegetation in the route corridor shall also be surveyed.

11.5.7 Pole (tower) location survey shall meet the following stipulations:

1 The pole (tower) location stake shall be set by adjacent control points, and its survey accuracy shall meet the requirements specified in Clause 1, Clause 4, Clause 5 of Article 11.5.5.

2 The following surveys shall be conducted during locating of the pole (tower):

    1) The centerline and sideline points with dangerous impacts.

    2) The location and elevation of objects, which are crossed over with dangerous impacts.

    3) If the transmission line goes through or approaches to slopes, steep shores and high buildings, the wind deviation cross – section or dangerous points shall be surveyed in accordance with design requirements.

    4) Straight line deviation degree and corner of route.

    5) The cross – section and topographic map of pole (tow-

er) base shall be surveyed if it is necessary for design.

3 The location of stakes or straight line stakes of pole (tower) shall be resurveyed before construction, and the following requirements shall be met:

    1) The allowable relative error of distance between stakes is 1/1000.

    2) The allowable comparative difference between resurveyed height difference and original results is 0.3m.

    3) The allowable comparative difference between resurveyed and original results of route straight line deviation degree and corner is 1'30".

4 The main technical requirements for surveying the overhead transmission line when the capacity is below 10kV can be lower properly. As for the overhead transmission line survey of 500kV or above, photogrammetry and GNSS method should be used.

11.5.8 Crossover survey shall meet the following stipulations:

1 If the route crosses a communication line, the upper line elevation of the center line intersection points shall be surveyed. The cross angle, the pole number on both sides, type and material shall be marked. If the centerline or side lines cross the pole (tower) top, the top elevation of pole (tower) shall be surveyed. If the elevation of left and right poles are different, the elevation of influential side line and wind deviation points shall also be surveyed.

2 If the route crosses or goes through power lines, the elevation of highest or lowest lines for center line cross points shall be surveyed. The cross angle, pole (tower) type and voltage class shall be marked. If the centerline or side lines cross the top of pole (tower), the elevation of pole (tower) top shall be surveyed. If the elevations of left and right of pole (towers) are different, the elevation of highest or lowest lines for the intersection points of influential side lines and that of wind deviation points shall also be surveyed.

3 If the route crosses railways and roads, the rail top and road elevation of cross point, road direction and mileage shall be surveyed. If the route crosses an electrified railway, line height and cross angle of intersection points shall also be surveyed.

4 If the route crosses or is close to houses, the elevation of the roof at the cross points or the distance and roof elevation of the location close to the house shall be surveyed.

5 If the route crosses rivers, reservoirs and inundated areas, the flood level and elevation of water level shall be surveyed according to the design requirements.

6 If the route crosses cableways, special pipelines and aqueducts, the top elevation of center line cross points shall be surveyed, and the name and material of crossed objects shall be marked.

11.5.9 For the survey of electric power transmission line, the following data shall be sorted out and submitted:

1 In the preliminary design stage, the following data shall be sorted out and submitted:

  1) Route scheme drawing.
  2) Plan and section map of important crossovers.
  3) Plan for incoming and outgoing lines of transformer station.
  4) Plan of crowded area.
  5) Relative location map of dangerous influences by the transmission line.

6) Technical summary report.

2 In construction documents design phase, the following data shall be sorted out and submitted:
1) Route path map.
2) Plan and section map of route.
3) Plan and section map of important crossovers.
4) Plan of incoming and outgoing lines of transformer station.
5) Plan of crowded area.
6) Relative location map of dangerous influences by the transmission line.
7) Section map of tower base.
8) The interval distance of straight line stakes and elevation result.
9) Technical summary report.

11.6　Road Survey

11.6.1 This section is applicable to the survey of special-purpose roads and railways in water conservancy and hydropower project construction. Road survey shall include initial survey and validation survey.

11.6.2 In the initial survey, route control survey, strip or regional topographic map survey and setting-out shall meet the following requirements:

1 Control survey shall comply with the following requirements:
  1) The road surveying control network should be laid out based on an overall consideration. More than two horizontal control points and two or three vertical control points shall be laid at large scale or complex junctions,

both sides of bridges or inlet and outlet of tunnels. If the bridge is shorter than 100m, only one height control point on one side of the bridge may be set.

2) The technical requirements for different class of horizontal and vertical control survey shall conform to the stipulations of Chapter 4 and Chapter 5.

2 Topographic mapping shall be in compliance with the following requirements:

1) The mapping scope of topographic maps shall be determined according to the highway classification, topographic condition and the design requirements. For the second-class road or above, the distance from the two sides to the center line shall be not less than 300m. When the onsite alignment method is adopted, the distance should not be less than 150m. If the expressway and the first-class highway have separated subgrade, the topographic map shall cover the intermediate belt; but if the two routes are far apart or the intermediate belt is rivers or mountains, the topographic map of the intermediate belt does not need to be surveyed. The scale of topographic map shall be 1:1000-1:5000.

2) The mapping scope of the bridge topographic map: 2-3 times the length of the bridge upstream and 1-2 times downstream; along the bridge axis, both sides shall be measured to the highest flood level or 2m above the design water level or 50m beyond the flood line. The scale of bridge topographic maps should be 1:500.

3) The mapping scope of tunnel topographic maps: in lat-

eral direction, 200m on both sides of the center line, and it can be widen appropriately based on the auxiliary project requirement or complicated geological situation; and in longitudinal direction, it shall be not less than 200m beyond the estimate excavation zero point, and separated tunnel shall be measured 100m beyond the monolithic subgrade meeting point. The mapping scale should be 1 : 2000, and 1 : 500 at the tunnel portal.

4) For the construction area with complex geology and large scale of protection project or important facility site, the 1 : 500 - 1 : 2000 topographic map shall be prepared.

5) The topographic mapping scope of interchange shall meet the requirements for the layout of interchange. Scale of 1 : 2000 should be adopted, but scale of 1 : 500 can be adopted if needed. If the large interchange with simple terrain and few land features, the 1 : 5000 can be adopted.

6) The topographic map for a large borrow area shall be surveyed at scale of 1 : 100 - 1 : 5000.

3 Route survey shall meet the following stipulations:

1) Aligning of second-class highway or above shall be carried out on the map after topographic survey while aligning of third-class highway or below may be carried out on site.

2) Aligning of vertical - section and cross-section in the map can be interpreted based on the topographic map. But for sections and places with strict elevation requirements, high fill and deep excavation sections,

large bridges, tunnels, interchanges, and areas which need special control, setting-out survey shall be carried out on site.

3) On-site route aligning may use the direct intersecting method or extending and setting turning point (or intersection points) method to determine the coordinates of intersection points and turning points. When the intersection or turning points are used as surveying control points, protection of stakes shall be carried out, and the angle and length between the selected points of intersection shall be surveyed according to the requirements of the fifth order horizontal control survey.

4) On site route setting-out may use RTK method, polar coordinate method, freely set station method, offset method and method of deflection angle, etc.

5) Route setting-out shall satisfy the stipulations specified in Table 11.6.2 when using freely set station method, offset method and deflection angle method.

Table 11.6.2 Closing error of centerline setting-out

Item	Express way, first-class and second-class Highway	Third-class highway and below
Angle closing error (″)	$30\sqrt{n}$	$60\sqrt{n}$
Relative closing error of length	1/2000	1/1000
Note: n is the number of break angles.		

6) The density of setting-out stakes in route aligning shall meet the requirements of survey and investigation. The location of setting-out stakes, the elevation of center stakes and the measurement accuracy

of cross – section shall comply with the stipulations concerning the center line surveying in the validation survey stage.

7) For large and medium sized bridges the axis and approach shall be set out on site, and vertical – section and cross – section shall be surveyed at the same time.

8) In the tunnel route aligning, the central line near the portal of the tunnel shall be set out onsite, and the vertical – section and cross – section of the tunnel portal shall be checked and surveyed.

9) For the survey of small bridge, over-flow bridge, complex culvert, ditch engineering, irrigation canal, stakes should be set out on site with measurement of elevation and section.

10) For interchange, grade separation, and complex at-grade intersection, intersection stakes shall be set out on site, height and section shall also be surveyed as necessary.

11) At the intersection of highway and railway, the rail track elevation and cross angle shall be surveyed; at the intersection of highway and pipeline, the intersection location, length, intersection angle, hanging height or embedment depth, pole (tower) height and affected length shall be surveyed.

11.6.3 In the validation survey stage, measurement shall include double-check of route control survey and supplementing survey or re-survey, revision survey and supplementing survey of topographic maps, center line layout, section survey, and it shall meet the following stipulations:

1 Control survey shall be in compliance with the following requirements:

1) Check the horizontal and vertical control survey of route being laid at the initial survey, and carry out supplementing survey if control points are destroyed or the scheme is changed without satisfying the requirements; if the discrepancy between the checking results and the initial survey results is over the limit, the whole control network shall be double-checked or re-surveyed.

2) If the accuracy of route control survey, distribution of control points and specifications of surveying marks cannot meet the construction requirements of the bridge and tunnel, the bridge and tunnel control network shall be set up, and the coordinate system shall be consistent with the route control survey. But if the length projection deformation of the route coordinate system affects the accuracy of the bridge and tunnel control survey, the independent coordinate system shall be adopted. Along the axis of the bridge, two or more bridge control stakes shall be set on each bank. And more than two inter-visible control points shall be set near the portal of the tunnel. The control points of extra-large bridges should adopt an observation pier with forced centering device. The main technical parameters of bridge control network survey shall be in compliance with stipulations specified in Table 11.6.3-1. The accuracy of tunnel control network shall be determined according to the whole length of the tunnel.

Table 11.6.3-1 Main technical parameters of bridge control network

Road class	Relative RMSE of the bridge axis	Horizontal level	Elevation level
Expressway, first-class highway	1/40000	Fourth class	Fourth class
Second-class highway and below	1/20000	Fifth class	Fifth class

Note: For the control network survey of oversized bridges, special design shall be carried out.

2 Center line survey shall meet the following stipulations:

1) The interval between center stakes of route shall not be greater than the values specified in Table 11.6.3-2. The starting and ending stakes and curve key stakes shall be set and surveyed accurately.

Table 11.6.3-2 Interval of center stakes   Unit: m

Straight line		Curves			
Plains, hills	Mountains	Curve with no super-elevation	$R>60$	$30<R<60$	$R<30$
50	25	25	20	10	5

Note: R is the radius of curve, m.

2) Stakes shall be added in special locations. The location and number of the supplementary stakes shall meet the investigation requirements of route, structures, facilities along the route, etc.

3) The discontinuous chainage stakes should be set at the straight line, but not in the scope of structures such as bridges, tunnels, interchange, etc. At the discontinuous chainage stakes, the mileage and increased or

decreased length shall be marked.

4) The position accuracy of the center stakes shall meet the stipulations specified in Table 11.6.3 - 3.

Table 11.6.3 - 3　Horizontal accuracy indexes of center stakes

Unit: m

Highway class	RMSE of center stake position		Detecting error of the pile position	
	Plains, hills	Mountains	Plains, hills	Mountains
Expressway, first-class and second-class highway	≤±50	≤±100	≤100	≤200
Third-class road and below	≤±100	≤±150	≤200	≤300

5) When the center line is set and surveyed by RTK method or polar coordinate method, it can be implemented in accordance with the fifth-order horizontal control survey stipulations specified in Chapter 4.

6) When the center stakes are surveyed and laid out using freely set station method, declination method, or offset method, its closing errors shall be smaller than the values stipulated in Table 11.6.3 - 4.

Table 11.6.3 - 4　Closing error of distance and angle measurement

Highway class	Longitudinal relative error		Lateral closing error (mm)		Closing error of angle (″)
	Plains, hills	Mountains	Plains, hills	Mountains	
Expressways, first-class and second-class highways	1/2000	1/1000	100	100	60
Third-class highway and below	1/1000	1/500	100	150	120

3 The elevation survey of center stakes shall meet the following requirements:

> 1) The elevation survey shall start and close at route elevation control points. Elevation shall be measured to the ground surface with the readings in millimeter, and the accuracy indexes shall meet the stipulations specified in Table 11.6.3 - 5.

Table 11.6.3 - 5 The survey accuracy of center stake height

Highway class	Closing error	Discrepancy between two measurements (mm)
Expressway, first-class and second-class highways	$\leqslant 30\sqrt{L}$	$\leqslant 50$
Third-class road and the below	$\leqslant 50\sqrt{L}$	$\leqslant 100$

Note: L is the route length in vertical survey, km.

> 2) For buildings, pipelines and rails, which need special control survey along the route, the elevation shall be surveyed in compliance with relevant stipulations, and the discrepancy between two measurements shall be less than 20mm.

4 Cross - section survey shall meet the following stipulations:

> 1) The width of cross - section shall meet the requirements of subgrade, drainage design, appendage setting-up, etc.
>
> 2) Cross - section survey shall be measured by stake, and the direction shall be perpendicular to the tangent of route center line. The reading precision of the distance and the elevation difference in cross - section measurement shall be up to 0.1m, and the test differences shall meet

the stipulations specified in Table 11.6.3-6.

Table 11.6.3-6   Tolerance of cross-section detection

Highway class	Distance	Elevation difference
Expressways, first-class and second-class highways	$S/100+0.1$	$h/100+S/200+0.1$
Third-class road and below	$S/50+0.1$	$h/50+S/100+0.1$
Note: S is the horizontal distance between the survey point and the center stake, m; h is the elevation difference between the survey point and the center stake, m.		

5 Topographic mapping shall meet the following stipulations:

1) The topographic map shall be checked on site in validation survey. Revision survey shall be conducted at sections, where terrain or land features changes. Supplementing survey shall be conducted if the scope of topographic maps cannot meet the design requirements. The whole topographic map shall be resurveyed if change is great on site.

2) The technical requirements of revision survey or supplementing survey for topographic maps shall meet the stipulations specified in Chapter 6.

11.6.4   For road survey project, the following data shall be sorted out and submitted:

1   Technical design document.
2   Control survey data.
3   Section survey data.
4   Vertical-section, cross-section and topographic maps.
5   Technical summary report.

11.7 Geological Exploration Survey

11.7.1 The following works shall be implemented in geological exploration survey (including geophysical prospecting survey):

    1 On-site setting out of the main geological investigation points according to the design position in the map, including the main geophysical prospecting line and points, etc.

    2 Measure the coordinates and elevations of building boreholes, shafts, adits, exploratory trenches, exploratory pits, geological points and geophysical prospecting points.

    3 Plot the connecting surveyed geological investigation points on the topographic map with a larger scale, and compile the distribution map of geological investigation points.

    4 Analyze and sort out the measurement achievements.

11.7.2 The conjunction survey accuracy for all kinds of geological investigation points shall meet the stipulation specified in Table 11.7.2.

11.7.3 Scales of the geological survey, base map, and setting-out conjunction survey shall meet the following stipulations:

    1 For geological survey at scale between 1 : 1000 and 1 : 10000, topographic maps at the same scale can be used as the base map, and the accuracy of the same scale topographic maps shall be used in setting-out and conjunction survey.

    2 For geological survey at scale 1 : 500, topographic map at double size enlargement of scale 1 : 1000 can be used as the base map, the accuracy of the 1 : 1000 topographic maps shall be used in setting-out and conjunction survey.

    3 When geological survey at two different scales is carried out in the same area, all kinds of geological investigation points may be implemented a conjunction survey according to the accuracy of the larger scale.

Table 11.7.2 Connecting survey accuracy of geological investigation points

Category of geological investigation points	Contents	RMSE of location relative to the adjacent mapping control points (mm in the map)		Vertical RMSE of relative to the adjacent basic vertical control points (m)	
		Plains, hills	Mountains, high mountains	Plains, hills	Mountains, high mountains
I	1. Borehole of hydraulic structure area; 2. Starting points of shaft, adit		±0.3	±0.06	±0.1
II	1. Borehole, wells, springs, and river water level sites used for groundwater table measurement; 2. Borehole used for groundwater dynamic observation	±0.75	±1.0	±0.06	±0.1
III	1. Pit, trough and geological points; 2. Geophysical prospecting point; 3. Section point; 4. Borehole at stock yard and general areas	±0.75	±1.0	Vertical RMSE relative to the adjacent measurement points (m)	
				±$\frac{1}{3}h$	±$\frac{1}{3}h$

Notes: 1. h is contour interval;
2. If the above connecting survey accuracy need be changed at individual geological investigation points, it shall be clarified in the task assignment.
3. Allowable setting-out error of borehole or other geological surveying points can be 1 time larger than the value stated in the table, and that of water borehole can be 1～2 times larger.
4. When horizontal scale of section map is larger than 1∶1000, measurement accuracy shall still comply with the scale of 1∶1000.
5. The setting-out accuracy of the opposite direction breakthrough flat adit shall be stipulated in the task assignment.

11.7.4 If geological survey is conducted before the photogrammetry, obvious surveying mark can be set on the geological investigation points before photography. If the aerophoto is available, the geological investigation points shall be annotated on it in field investigation and plotting. When mapping in office, the geological investigation points shall be surveyed on the topographic map. And their coordinates and elevation may be surveyed from the photo by the photogrammetry method when necessary.

11.7.5 Before setting out the boreholes, well-holes, pothole, and the main geophysical prospecting line, the parameters such as the azimuth angle and distance to the known control points shall be calculated according to the coordinates.

11.7.6 The elevation survey of geological exploration shall meet the following stipulations:

1 The elevation of class I and class II geological points shall be connecting surveyed with mapping control leveling. And the intersection trigonometric leveling method or GNSS method for the elevation surveying may be used if necessary. However, the discrepancy of elevation measured by intersecting from three directions or two GNSS measurements shall not be larger than 0.3m.

2 The elevation of class III points shall be surveyed by two different/respective observation stations using the elevation note method of terrain mapping.

11.7.7 The accuracy for coordinates and elevation of all kinds of geological investigation points shall meet the stipulation of Table 11.7.7.

11.7.8 In geological investigation survey, if no control network available or the density of original control points cannot satisfy the requirement, the basic control shall be laid out according to the stipulation specified in Chapter 4 and Chapter 5.

Table 11.7.7 Coordinate, elevation accuracy of geological investigation points Unit: m

Category of geological investigation points	The specific contents of geological investigation points	Coordinates		Elevation	
		Measurement and calculation	Final result	Measurement and calculation	Final result
I	1. Borehole of hydraulic structure area; 2. Starting points of shaft, adit	0.001	0.01	0.001	0.01
II	1. Borehole, wells, springs, and river water level sites used for groundwater table measurement; 2. Borehole used for groundwater dynamic observation	0.01	0.1	0.01	0.01
III	1. Pit, trough and geological points; 2. Geophysical prospecting point; 3. Section point; 4. Borehole at stock yard and general areas	0.01	0.1	0.01	0.01

11.7.9 Near the adit or shaft, not less than two inter-visible fixed mapping control points and more than one of fifth-grade elevation points shall be set.

11.7.10 The setting out of boreholes shall meet the following stipulations:

   1 The borehole shall be set out according to the design of the hole position in the map or the geologists' requirements.

   2 The setting out goal of ground borehole shall be the orifice center, and that of water borehole shall be the drill shaft or the top of derrick.

   3 After the setting-out of borehole, a directional stake shall be set outside the scope of the machine, and a sketch shall be drawn on the field handbook, indicating the distance and azimuth from directional stake to the borehole.

11.7.11 Connecting survey of the boreholes shall meet the following stipulations:

   1 For ground boreholes, the coordinates, ground elevation and the pipe mouth elevation of the actual borehole center shall be surveyed.

   2 For water borehole, the coordinates, ground elevation of the pipe mouth and the river bed elevation at the borehole shall be surveyed.

   3 Near the borehole or well used for groundwater dynamic observation, the fifth order leveling monuments shall be laid out.

11.7.12 For well and shaft survey, measurement of the shaft, one-way adit and one-way shaft hole shall be carried out, and it shall be implemented with the following requirements:

   1 The setting out and connecting survey of well and shaft shall meet the following stipulations:

1) The geometric center shall be marked on-site in setting out of shaft entrance. When connecting survey is implemented, the horizontal position and elevation of the longer diagonal or the two ends of the shaft entrance's longer side shall be measured for larger noncircular shaft; for smaller noncircular shaft, only the horizontal position and elevation of the wellhead's one angle need to be surveyed. The length of each side of the entrance and the position of connecting survey points shall be recorded in the handbook. For circular shaft, connecting survey of the horizontal position and elevation of both ends of the diameter may be carried out.

2) For the setting out of one-way adit, the center location and elevation of the adit bottom and the tunneling direction line shall be marked on-site. In connecting survey, the adit's center location and bottom elevation, and the adit's height, width, length and the azimuth angle of axis shall be measured.

3) For the setting-out of one-way shaft hole wellhead, the geometric center of the entrance shall be marked on-site, and its longer diagonal shall be consistent with the direction line of the adit. On either adjacent side of the shaft, the position of shaft center line shall be measured, and corresponding elevation shall be marked at every 5 - 10m. When bottom elevation of the adit has been reached, the direction line and elevation shall be transimited - into the adit accurately from ground. In the adit, its geometric center line shall be projected on the ceiling, and the surveying mark of

the midline points shall be laid in a properly kept place; and also the waist line shall be plotted on the shaft walls with 1m above the bottom. For connecting survey, besides the location and elevation of entrance, those of the turning points and ending points of the adit, and the width, height of the adit and other data shall be measured.

4) In the handbook and the result table, the plane and vertical - section map of the shaft and adit shall be drawn, and their measuring position, elevation and the relative size of geometric shapes in well hole shall be marked.

2 The horizontal control points on the ground shall be connected to those in the shaft by shaft plumbing. Heavy hammer plumbing method or laser plumbing instrument may be adopted for plumbing.

11.7.13 For connecting survey of the shaft and adit, the relationship between horizontal coordinates and elevation of the ground control points and those of the control points inside the adit shall be measured, and the following stipulations shall be met:

1 When the connecting survey is implemented between the plane control points on the ground and those inside the adit, the connecting triangle method or connecting direction line method may be used.

2 The adit orientation of one-way shaft adit may use reversal point method or transit method by gyro theodolite which has a RMSE of less than $\pm 60''$ for the single-pass measurement.

3 For the orientation or connection survey, instruments and survey targets shall be centered, and the allowable centering error is

1 mm. For survey on the ground and underground, the instrument shall directly sight the netsuke line. And the survey shall be carried out when the instrument is motionless and stable.

4 Shaft plumbing, orientation and connecting survey shall be carried out twice separately adopting different methods, different graphics and different routes by different personnel; the permissible discrepancy between azimuths of the first traverse in hole twice measured shall be 5″.

5 Ground elevations may be transmitted through the shaft to the inside of tunnel by the following methods:

  1) The long steel ruler is used to directly import elevation once.

  2) The short steel ruler is used to import elevation constantly in sections.

  3) The long steel wire is used to directly import elevation once.

  4) The electroptic range finder is used to directly import elevation.

6 When starting adit excavation, total station or theodolite may be used for guiding. If the adit is long, laser guide instrument or laser directional total station should be used for guiding.

11. 7. 14 Survey of pit and trough shall comply with the following stipulations:

1 For trial pit, the surface elevation and plane position of the two endpoints of a diagonal or the longer side shall be measured, and a sketch shall be drawn in the handbook with their corresponding position and side length being marked.

2 For exploratory trench, the plane position of its endpoints and turning points, the elevation of surface and the trench bottom along one side shall be measured, and a sketch shall be drawn in the handbook with its corresponding position and width being marked.

3 For the large pit or trench in maps, the boundary may be drawn using symbols of land class boundary with their names also be marked.

11.7.15 All kinds of geological points shall be surveyed and mapped in the topographic map, and it shall meet the following stipulations:

1 For geological mapping at a scale no greater than 1 : 1000, common geological points may be surveyed and mapped using visual method, compass intersection method, or handheld GNSS instrument. For the geological points with higher accuracy requirement, survey with instruments should be used.

2 For geological mapping at a scale between 1 : 1000 and 1 : 5000, survey with instruments should be used.

3 RTK method, total station polar coordinate method may be used for the geological point survey.

11.7.16 Connecting survey of geophysical prospecting points shall comply with stipulations specified in Table 11.7.16.

Table 11.7.16 The connecting survey of geophysical prospecting points and the achievement that should be advanced

Items	Names of connecting survey	Outcomes
Electrical prospecting	Electrical sounding points, section endpoint and turning points, the ground slope turning points, basis charging points, main abnormal points	Coordinates and elevations

Table 11.7.16 (Continued)

Items	Names of connecting survey	Outcomes
Seismic prospecting	Low-speed electric center points, section end points and the turning points, shot points	Coordinates and elevations
Microgravity prospecting	Survey points	Elevations
Radioactive prospecting	Survey points	Elevations
Hydro-acoustic exploration	Section end points on both sides of the cross	Coordinates, elevation and distance
Elastic wave testing	Shocking points	Coordinates and elevations

11.7.17 Main geophysical prospecting lines and points shall be set out on-site according to the design positions, and be marked with wooden stakes. After geophysical prospecting operations are carried out, connecting survey shall be carried out.

11.7.18 The accuracy and mapping scale of vertical - section measurement and shall meet the following stipulations:

1 The survey scale of geological section or geophysical prospecting section in hydraulic structure area may use a scale of 1∶1000 or 1∶2000. For other areas, the section scale shall be consistent with the project requirements.

2 Horizontal and vertical mapping scale of geological section or geophysical prospecting section may be determined by the length of the section and the relative elevation. The same section should not be drawn in two maps.

11.7.19 Position of vertical - section and monuments shall meet the following stipulations:

1 Vertical-section lines shall be marked on the topographic map firstly, and then the geological personnel, geophysical prospecting personnel, and survey personnel will mark the endpoints, folding points and vertical - section intersection points on site together.

2 Fixed monuments shall be laid at endpoints and turning points of important section in hydraulic structure area. The specification of monuments may follow the requirements of the monument construction.

11.7.20 Basis points and turning points of vertical - section shall meet the following stipulations:

1 The allowable RMSE of plane basis points shall be 0.2 mm in the map with the horizontal scale of actual accuracy section. And the permissible elevation RMSE of elevation basis points shall be 1/10 of the basic contour interval.

2 Plane and elevation basis points should be laid on the same points and set on the endpoints of the vertical - section.

3 The allowable RMSE for plane position of turning points shall be 0.3mm in the map with the horizontal scale of actual accuracy section. The permissible RMSE of elevation shall be 1/6 of the basic contour interval. The turning points of broken line vertical - section shall be measured according to the accuracy of turning points.

4 At least one basic point shall be set on one vertical - section. If the vertical - section is longer than 1000m, one basic point should be respectively set at each end of the vertical - section.

5 The elevation survey of turning points shall start from elevation basic points and close at leveling points.

11.7.21 The following methods may be used for the vertical -

section point collection:
1 Total station survey method.
2 GNSS survey method.
3 The photogrammetry method is used to measure vertical - section on the photograph.
4 Use the topographic map with the large scale to truncate vertical - section.

11.7.22 The density of vertical - section points shall be one point every 1 - 3 cm in the map. The horizontal position and elevation of topography changing points, land feature points, geological investigation points, intersection points of different vertical - sections, the water's edge points, etc. shall be surveyed.

11.7.23 The following do cuments shall be sorted out and submitted for geological exploration survey:
1 The handbook of survey records and calculation data.
2 The result table of the horizontal, vertical control points and result tables of vertical - section.
3 The result table of geological survey points.
4 The distribution map of geological survey points.
5 Vertical - section.
6 Technical summary report (if necessary).

11.8 Survey in Water Area

11.8.1 Water area survey shall be in compliance with the following general requirements:
1 The scale of underwater topographic maps for water conservancy and hydropower project investigation and design shall be chosen according to the stipulation set out in Table 11.8.1 - 1.

Table 11.8.1-1 Scale for underwater topographic maps

Stage	Survey area	Mapping scale	Remark
Planning	Rivers, sea area	1:2000-1:10000	Every five years
Feasibility study	Reservoir, hydropower station, sluice and dam project	1:1000 - 1:2000	Project region
		1:1000 - 1:5000	Non-project region
	Water diversion project	1:1000 - 1:2000	
	River regulation project	1:1000 - 1:2000	
Preliminary design	Reservoir, hydropower station, sluice and dam project	1:500 - 1:1000	Project region
		1:1000 - 1:2000	No-project region
	Water diversion project	1:500 - 1:1000	
	River regulation project	1:500 - 1:1000	
Construction design	Hydraulic structure area	1:200 - 1:500	
	Water diversion project	1:200 - 1:500	Important building area
		1:500 - 1:1000	
	River regulation	1:200 - 1:1000	
Operations management	Reservoir, hydropower station, sluice and dam project	1:500 - 1:2000	Once every 2-3 years
	Water diversion project	1:500 - 1:2000	Once every 2-3 years
	River regulation project	1:500 - 1:2000	Once every 2-3 years

Note: Water diversion project includes channel, canal, pipeline; River regulation project includes river embankment and sea dike.

2 Before starting the work, relevant information of measurement area shall be collected, and if necessary, field investigation shall also be carried out.

3 For underwater topographic survey, the survey plan shall be made according to the underwater terrain conditions such as the water depth, flow velocity, flow pattern, river width, river bed, tidal, sea surface and the proposed instruments and equipment, and the survey shall be conducted in an appropriate season.

4 Bathymetric survey may adopt the conventional mode and the automatic mode:

1) The conventional mode may use intersection method, polar coordinate method and section wire method for plane positioning, and the sounding rod, sounding lead and simulation sounder for bathymetric survey.

2) The automatic mode may use GNSS method for plane positioning, and digital single beam depth finder or the multi-beam echo sounder for bathymetric survey, and the two kinds of data shall be simultaneously acquired.

5 Water level (the elevation of water surface) shall be measured by gauging station (tide staff), and the local project may be measured directly. The fixed tide staff of long-term gauging station shall meet the requirements of GB/T 50138, and its elevation may be connecting surveyed using the fifth class or the higher class.

6 If the water level difference at two banks of river is more than 0.1m, the traverse gradient correction shall be carried out. If the water surface of river center changes abnormally affecting the precision of water depth, river center gradient shall be measured additionally.

7 If the survey area is far from the shore and the water level difference between survey area and the shore is larger than the allowable, temporary gauging station may be set on the sea.

8 The limits for the location RMSE of sounding positioning points shall comply with the requirements specified in Table 11.8.1-2.

9 The limits for the depth RMSE of the sounding points shall comply with the requirements set out in Table 11.8.1-3.

Table 11.8.1-2 **The limits for the location RMSE of sounding positioning points**

Mapping scale	The limits for RMSE of location (mm on map)
1:500 - 1:2000	1.5
1:5000 - 1:10000	1.0

Table 11.8.1-3 **The limits for the depth RMSE of the sounding points**

Water depth	Sounder or tools	Flow velocity (m/s)	RMSE of depth (m)
0 - 3	Sounding pole	—	±0.10
0 - 10	Sounding lead	<1	±0.15
1 - 10	Depth finder	—	±0.15
10 - 20	Depth finder or sounding lead	<0.5	±0.20
>20	Depth finder	—	±0.015H

Notes: 1. H is the depth of water, m.
2. If the accuracy is not high or the measurement is difficult, the depth RMSE of the sounding points may relax as 2 times.

10 Water area survey should be carried out when the wave is smaller. If the wave height is over 0.6m in sea area and over

0.4m inland, the measurement shall stop.

11 The operation mode and the mapping scale shall comply with the stipulation set out in Table 11.8.1-4.

Table 11.8.1-4 The operation mode and mapping scale

Operation mode	Positioning method	Mapping scale	Notes
Normal mode	Ordinary method	$\leqslant 1:2000$	Local
Automatic changing mode	RBN GNSS SBAS GNSS	$\leqslant 1:2000$	
	GNSS RTD	$\leqslant 1:1000$	
	GNSS RTK	$\leqslant 1:200$	No tidal observation

12 A fixed cross-section should be laid every 2-5km, and the cross-section may be appropriately densified for important river section. It shall be measured more than one time each year.

13 The interval of cross-section shall meet the stipulations specified in Table 11.8.1-5. The water point interval of vertical-section shall follow the stipulations of cross-section.

Table 11.8.1-5 Interval of cross-section

Stage	Survey area	Survey area condition	Cross-section interval (m)
Planning	River	Mountains section	2-3
		Plain section	2-5
Feasibility study	Reservoir, hydropower station, sluice and dam project	Backwater reach	1-3
		Normal flow reach	2-5
	Water division project	Mountains section	0.2-0.5
		Plain section	0.2-1
	River regulation project	Mountains section	0.2-0.3
		Plain section	0.2-0.5

Table 11.8.1-5 (Continued)

Stage	Survey area	Survey area condition	Cross-section interval (m)
Preliminary design	Reservoir, hydropower station, sluice and dam project	Backwater reach	0.5-1
		Normal flow reach	1-2
	Water diversion project	Mountains section	0.05-0.1
		Plain section	0.05-0.2
	River regulation project	Mountains section	0.05-0.1
		Plain section	0.05-0.2
Construction design	Hydraulic structure area		0.01-0.03
	Water diversion project		0.02-0.05
	River regulation project		0.02-0.05
Operations management	Reservoir, hydropower station, sluice and dam project	Dam area	0.01-0.03
		Reservoir area	0.2-1
	Water diversion project	Building section	0.02-0.05
		General section	0.05-0.1
	River regulation project	Dangerous section	0.02-0.05
		General section	0.05-0.1

14 For special research section, one should be set every 0.2-1km.

15 Survey and mapping scale of the vertical-section and cross-section shall be determined according to the mapping content, section length, survey point density and section use. The selection shall comply with the stipulation specified in Table 11.8.1-6.

16 The allowable RMSE of horizontal position for land section survey is 0.8mm on map. The location accuracy of the

underwater section survey shall not be lower than that of 1 :
2000 cartographic maps.

Table 11. 8. 1 - 6 Survey and mapping scale of vertical - section and cross-section

Section category	Stage	Map category	Horizontal scale	Vertical scale
Fixed section		Vertical - section	1 : 50000 - 1 : 200000	1 : 100 - 1 : 1000
		Cross - section	1 : 500 - 1 : 5000	1 : 100 - 1 : 200
The survey design section	Planning, feasibility study	Vertical - section	1 : 25000 - 1 : 200000	1 : 100 - 1 : 1000
		Cross - section	1 : 200 - 1 : 2000	1 : 100 - 1 : 200
	Preliminary design	Vertical - section	1 : 10000 - 1 : 100000	1 : 100 - 1 : 500
		Cross section	1 : 200 - 1 : 2000	1 : 100 - 1 : 200
	Construction design, operations management	Vertical - section	1 : 2000 - 1 : 25000	1 : 100 - 1 : 200
		Cross - section	1 : 100 - 1 : 500	1 : 100 - 1 : 200
Special research section		Vertical - section	1 : 10000 - 1 : 100000	1 : 100 - 1 : 500
		Cross - section	1 : 200 - 1 : 2000	1 : 100 - 1 : 200

Note: Provided M is the horizontal scale, and when $(1/M) \times$ cross - section interval roughly equals to 1cm in the map the M shall be appropriate.

17 The survey accuracy of cross - section points shall meet the relevant stipulation of terrain points specified in Clause 4 and Clause 5 of Article 3. 0. 5.

11. 8. 2 The plan of horizontal and vertical control for water area survey shall be designed according to the conditions such as survey area size, task features, mapping scale and operation pattern, etc. Horizontal and vertical control network survey shall comply with the requirements set out in Chapter 4 and Chapter 5.

11.8.3 For water area survey, water level and elevation of water surface shall be surveyed in compliance with the following requirements:

    1 The layout of gauging station or tide staff shall meet the following stipulations:

        1) The number of gauging station shall be determined according to the operation mode. Gauging station shall be set in a place without the influence of backflow, backwater, the wind and waves and rapid flow shock. The obverse side of tide staff shall be parallel to the shore, in order to fully reflect the water level changing.

        2) Tidal waters area survey may use the hydrometric station near the survey area or set tide staff. The maximum space between tide staffs should not be more than 20km for tidal water area, and not more than 10km for tidal reach in river.

        3) Around (or both sides of) the lakes or reservoirs, tide staffs shall be set.

        4) For general river reach one tide staff shall be set up every 20km, while for other river reach with mountainous valley, complex river bed and rapids, more tide staffs shall be added.

        5) If the water level difference between the two river sides is more than 0.1m, tide staff shall be set up on both sides.

6) When the survey area is far from the shore and the observed water level data of shore can't fully reflect the water level of the survey area, more tide staffs shall be added.

2 The observation of water level or the surface elevation shall meet the following stipulations:

1) For direct determining method of water surface elevation height, level or total station may be used, and the leveling grade shall not be less than the fifth class.

2) Connecting survey for elevation of tide staff zero point shall not be lower than the fifth class leveling. Tide staff zero point shall be checked regularly. The tide staff shall be corrected immediately when it tilts, and its zero point elevation shall be checked. The hydrograph zero point shall also be checked frequently.

3) For water level observation, Chinese Standard Time shall be adopted, and water levels shall be synchronously recorded. The water level observation shall accurate to 10mm. If the water level difference between the upper and lower slope section is less than 0.2m, the accuracy of slope water level shall accurate to 5mm. The frequency of water level observation shall meet the stipulations specified in Table 11.8.3. The location of estuaries, coastal waters shall be observed once every 10 mins.

4) The automatically recorded data of reservoirs and lakes form local water conservancy administrative department may be used, but shall be checked.

Table 11.8.3　The frequency of water level observation

Area	Water level changing characteristics	Frequency of observation	Densified frequency
Sea area	Tidal influence	Once every 10 - 30 min	Once every 10 min
Inland water area	$\Delta H < 0.1m$	Once at the starting and the ending of the observation	
	$0.1m \leqslant \Delta H \leqslant 0.3m$	Once at the starting, the middle and the ending of the observation	
	$\Delta H > 0.3m$	Once every 1h	

Note: ΔH is the water level changes daily, m.

    5) If the difference of water level correction between the two tide staff located on upstream and downstream of nontidal river is less than 0.1m, their average shall be used as the water level correction; If the difference value is not less than 0.1m, correction in sections shall be calculated according to the linear interpolation method. For sea area, water level correction shall be calculated by linear interpolation, time difference interpolation method, regression interpolation, zoning interpolation method and other methods according to the water level changing.

11.8.4　Underwater topographic survey shall include the positioning survey and bathymetric survey, and shall comply with the following requirements:

    1　Sounding line layout shall meet the following requirements:

1) Sounding plan line should be vertical to the channel center line or coast line, and parallel to the main dam axis of the reservoir or lakes. Sounding plan line may be laid into a bunch of parallel lines, a spiral or a 45° oblique line.

2) The space between sounding plan lines shall meet the stipulation specified in Table 11.8.4-1.

Table 11.8.4-1　The space between sounding plan lines

Operation area	General waters (mm on map)	Important waters (mm on map)
Rivers, lakes and reservoirs	15 – 30	10 – 20
Artificial canals, irrigation channels	15 – 25	8 – 15
Estuaries, the coast waters	10 – 25	8 – 15

3) Sounding check line should be perpendicular to the sounding line, and its length should not be less than 5% of the total sounding line. For adjacent segments, one overlapping sounding line shall be laid, and for the adjacent section in different periods, two overlapping sounding line shall be set.

2　Positioning survey shall satisfy the following requirements:

1) The space of sounding positioning points shall be 10 – 30mm on the map, for area such as underwater complex terrain or important water areas, the measuring points should be more.

2) Positioning center shall be consistent with the sounding center, and the deviation should not be greater than 0.3mm on the map, eccentricity correction shall be carried out when the deviation is

transfinite.

3) If the intersection method and polar coordinates method are used for positioning, starting direction shall be checked during and before the ending of operation, and the tolerance is $1''$; the scope of intersection angle measuring by intersection positioning method should be controlled between $30°-150°$.

4) When using section wire method for positioning, the relative tolerance of the wire length shall be less than 1/200.

5) When GNSS method for positioning is adopted, the antenna of mobile station shall be firmly placed in a higher place of broadside and insulated from metal object. The antenna position should be in the same vertical with sounder transducer, and eccentricity correction shall be carried out when the deviation is greater than 0.3mm on the map. The relevant stipulation of the GNSS positioning is referred to in Article E.2.1 in Annex E.

6) Positioning data and sounding data shall be synchronously measured, otherwise delay correction shall be implemented.

3 Bathymetric survey shall satisfy the following requirements:

1) The time of survey ship, gauging station and positioning station shall be checked before the starting of the measurement. Water level observation shall be started 10 min before the survey starts, and finished 10 min after the survey ends.

2) For the sounding, the analog recording digital sound-

er should be used, and sounder inspection shall meet requirements in Article E.2.2 of Annex E; sounder transducer should be installed at 1/3 - 1/2 of ship's length from the bow, static deep-draft should be 0.3 - 0.8m, and the accuracy of installation is 1cm; when powerboat is used to sound, dynamic draft correction of sounding transducer shall be measured according needs, and the measuring methods is referred to in Article E.2.3 in Annex E; sound velocity correction shall be calculated for sea area measurement, and for shallow water with depth less than 20m, its correction may be calculated using known depth or sounding board; in addition, sound velocity of water measuring area shall be measured using the sound velocity meter, the sound velocity of sounder shall be set according to the average velocity of sound in measuring area. In order to ensure the accuracy of bathymetric survey, the depth shall be compared with the depth of inspection plate.

3) Sounding and testing in water area where the depth change fast should be arranged on the same day, and at the intersection of check line and sounding line, the differences of water depth between the sounding line and testing line within the scope of 1mm on the map shall meet the stipulation specified in Table 11.8.4-2.

Table 11.8.4-2 Difference tolerance of water depth testing

Unit: m

Water depth	Difference Tolerance
$H \leqslant 20$	$\leqslant 0.4$
$H > 20$	$\leqslant 0.02H$

4) If the traditional tools for depth sounding are adopted, sounding rod with division of 0.1m is suitable for the survey area with the flow velocity less than 1m/s, and water depth less than 5m; hand cast sounding lead is suitable for survey area with the flow velocity less than 1m/s, and water depth less than 10m. Sounding-line with small retractility and good tensile strength should be chosen, and the allowable error is 1/100 of the sounding-line length; the fish lead or other heavy hammer for sounding should be 15 – 50kg.

4 Wave corrections shall meet the following requirements:

1) If the GPS-RTK for positioning is adopted, the wave correction of sounding results may be cancelled.

2) If the multi-beam sounding system for the water depth measurement is adopted, the attitude correction of depth measurement results shall be calculated automatically in the measuring post-processing.

3) If the single beam sounding use wave compensator for wave correction, wave compensator should be close to the sounder transducer; sounder or data acquisition software should record the original sounding data, wave data and water depth correction data simultaneously; wave compensator shall not be moved in the process of measuring.

5 Supplementing survey and resurveying of sounding shall comply with the following requirements:

1) Sounding shall be supplementing surveyed at any kind of situations where the sounding line space is more than half of the space given in Table 11.8.4 – 1, sounder zero signal is abnormal, the echo signal is in-

terrupted or is ambiguous, two or more connecting anchor points are missed, the starting, ending and the turning points of the line are not be positioned, sounding points mark is not accordance with the position points mark and unable to be corrected.

2) Sounding shall be resurveyed at any kind of situations where sounding device is abnormal, the underwater topography is unreasonable, GNSS positioning is abnormal and accuracy of self-checking is unqualified, mean square error of positioning is transfinite, RMSE of depth is transfinite or the depth compared points is over 20% of the total compared points.

6 Using multi-beam sounding system for waters sweeping survey, the following requirements shall be meet:

1) The scale of the multi-beam complete coverage sweeping survey should be 1 : 1000 – 1 : 2000 for rivers, reservoirs or lakes, while 1 : 1000 – 1 : 5000 for sea area.

2) Coverage shall be overlapped between sweeping survey lines; if partition sweeping survey is carried out, the overlapped area of two adjacent measurement areas shall be more than 5%.

3) The width of sweeping survey using multi-beam sounding system is 3 to 8 times of the water depth, which is recommended for rivers, reservoirs and lakes water survey in the wet season. And it should not be used for the area that depth less than 3m.

4) The requirements for multi-beam sounding system sweeping survey are specified in Article E. 2. 4 of Annex E.

7 For dangerous river section, reservoir silting – up area and important water conservancy area, regular sweeping survey should be conducted once a year or more.

11.8.5 For sounding survey, the following office work shall be done:

1 After input water level observation data and other correction data by sounding software, water level, dynamic water deep draft and wave correction shall be calculated automatically. For conventional sounding conducted, correction shall be calculated by hand.

2 Check the rationality of the sounding data and delete the wrong sounding points. The leapfrog points shall be checked and corrected when using GPS-RTK positioning. If the leapfrog points cannot be corrected they shall be removed. But the number of removed points shall not be more than 10% of the total points.

3 Choose the sounding line according to the tracking map, and input it into the computer.

4 Make plan of supplementary survey and resurvey.

5 Check and organize all field handbooks, and bind them together.

11.8.6 Mapping of underwater contour or depth contour shall meet the following requirements:

1 Use sounding survey software to calculate the final sounding results the mapping height datum, and save the elevation results of underwater surveying points.

2 Input shoreline survey data and underwater elevation data into the topographic map software system, and use the software to generate the underwater digital terrain model (DTM).

3 Using the topographic map software generate underwater contour automatically, and finish the mapping of underwater topographic map by editing, modification and smoothing.

4 According to different requirements, sounding result may be converted to the depth datum, the average sea surface or sailing datum. The final result of underwater sounding points is the depth. Draw depth contours based on the depth of sounding points.

11.8.7 The merging, processing and mapping for lands and water topography shall meet the following requirements:

 1 If underwater terrain and land terrain are merged into one map, the commissure of the shoreline and land terrain shall have 4mm overlapped; The allowable error of land feature at the commissure is $2\sqrt{2}$ times than the permissible RMSE of land feature; The permissible error of contour at the commissure is $2\sqrt{2}$ times than the permissible RMSE of elevation; the true configuration of land feature at the commissure shall not be changed, and stitching of the geomorphy shall not lead to deformation.

 2 Land contours and water boundary lines shall be plotted with solid lines, and the underwater contour lines shall be plotted with the blue solid line.

 3 Stipulation for the space of underwater spotting elevation point: 1cm to 2cm on the map for small and medium scale, 1cm to 3cm on the map for large scale, and more underwater spot elevation point may be required at important place.

 4 The subdivision of underwater topographic map may refer to the subdivision of land topographic map. Edit and decorate by map block firstly, then merge blocks to a whole map and then subdivide it. Finally output the drawings in subdivisions.

11.8.8 Cross-sectional survey shall comply with the following stipulations:

 1 The location of fixed cross - section for planning, feasibility study and special research shall meet the following requirements:

  1) Lay out the cross - section on topographic map at medium or small scale, and then survey and select the

cross‐section on‐site.

2) Cross‐section shall be set in the locations such as the cross‐section or length changing significantly, convergence of tributaries, sharp turns in the rivers, slope changing obviously, villages and towns, industrial and mining enterprises, cultural relics, landslide, temporary water gauge, etc.

3) Cross‐section shall not be set in dangerous shoals, rapids or vortex area.

4) River cross‐section shall be perpendicular to channel, and reservoir cross section shall be perpendicular to the center line of the reservoir.

5) At the cross‐section endpoints of the dam site and the sluice gate site, permanent surveying marks shall be laid.

2 Cross‐section for the preliminary design, construction design and operation management shall be laid out on the topographic map of large or medium scale firstly, and then determined the location on site roughly using positioning equipment such as GNSS receiver, total station, etc. Cross‐sections shall be set at the turning points of water course, channel, levee and hydraulic structure such as bridge, gate dam, crossing-dyke sluice and culvert.

3 Cross‐section stakes shall divide into permanent, normal and temporary surveying marks. Permanent surveying marks shall use class control point monument, normal surveying marks shall use mapping control point monument, and temporary surveying marks shall use stake or chisel stone.

4 Elevation of cross‐section stakes shall be surveyed by the fifth class leveling or better, and may be surveyed using GNSS elevation method or trigonometric leveling at difficult areas. And the accuracy shall not be less than the map accuracy.

5 Embedment and measurement for all kinds of cross-section stakes shall meet the stipulations specified in Table 11.8.8.

Table 11.8.8 **Requirements of embedment and survey for cross-section stakes**

Classification	Stages	Cross-section stake surveying mark	The number of section stake	Accuracy of horizontal control
Fixed section		Permanent	Each on both sides of the section	Fifth class
The section for survey and design	Planning	Permanent	Each on both sides of the section	Fifth class
	Feasibility study	Permanent or ordinary	Each on both sides of the section	Fifth class or mapping
	Preliminary design	Ordinary or temporarily	Each on both sides of the section or one stake one azimuth	Mapping
	Construction design	Ordinary or temporarily	Each on both sides of the section or one stake one azimuth	Mapping
	Operations management	Permanent	One on both sides of the section	Fifth class
Special research section		Ordinary	One on both sides of the section	Mapping

Note: The section stakes on both sides shall be set on the two sides of the river. When the back and forth stakes are laid out on one river bank, the space of the two stakes shall not be less than 1/10 the length of the section; If only one section stake is set, the azimuth of the section line shall be measured.

6 Cross-section may be measured with traditional or automatic mode using the combination of GNSS measurement system, total station and sounder, sounding tools.

7 Deviate distance between the measured line and section line in the cross-section survey is 5mm on the map for the water area mapping at scale of 1 : 200 – 1 : 1000, and 3mm on the map for the water area mapping in scale of 1 : 2000 –1 : 5000, and is 2mm on the map for land.

8 The space of cross-section points is 1 – 3cm on the map for land part, and 0.5 – 1.5cm on the map underwater. If the river channel is narrow, the sounding points shall not be less than 3.

9 The two ends of the land feature contour, slope transformation location, water edge and underwater thalweg points shall be measured. When the roughness of subsurface need to be studied, demarcation points of terrestrial vegetation and soil shall also be measured, and the properties shall be noted.

10 The water area of fixed sections, monitoring section of the reservoir and river dangerous section shall be observed directly and reverse, the discrepancy of the water area between the two surveys shall be less than 2%. If the discrepancy is in the allowable range, the average value of the two surveys or a walk line closest to the section line shall be used for the section map.

11 If the RTK and digital sounder for automatic survey are adopted, the elevation survey of water surface may be skipped, and the 3D data (X, Y, H) of section points shall be acquired directly.

12 The water surface elevation may be determinated by leveling method or EDM trigonometric leveling method. If the water level fluctuates greatly, the water levels of both sides shall be observed simultaneously. Measure the water level at starting, middle and ending of the section survey, and the average of the observation result shall be the water level of section survey.

　　13　If RTK or total station is used for the cross-section survey of land part, the 3D data (X, Y, H) of section points may be acquired directly.

　　14　After cross-section survey is completed, water and land measurement results shall be converted into 3D data with unified starting point distance or unified measurement datum, and for water and land part there shall be 1-2 overlapped points.

11.8.9　Vertical-section survey shall comply with the following stipulations:

　　1　Vertical-section points and thalweg points may be measured on-site or use cross-section drawing to select points, and the points in changing riverbed shall be densified and surveyed on site. In the stages of planning and feasibility study, points of vertical-section and cross-section may be selected from new measured underwater topographic map. The mileage of vertical-section may be acquired from coordinate data of cross-section, or through measuring on the topographic map whose scale is not less than 1:10000.

　　2　The vertical-section water level points including thalweg points shall be selected from the following parts:

　　　　1) The upstream and downstream of waterfalls, water drop and rapids.

　　　　2) Cross-section, tide staff and tributary confluence.

　　　　3) Inlet and outlet of tunnel, dam site, gate and bridge.

　　3　For the vertical-section map with long river reach, small gradient and limited measurement period, the water line at the same time shall be adopted. For vertical-section map with large gradient, small water level changing, short mileage and short measurement period, the working water level line may be used as water line at the same time directly.

4 Elevation of the identified historical flooded marker shall be connecting surveyed with the accuracy above the mapping leveling; the coordinates of traces points may be measured by handheld GNSS or measured on topographic map. Low water level, normal flood level may be calculated from hydrological data.

11.8.10 For water measuring project, the following documents shall be collated and submitted:

1 Technical design document.

2 Topographic maps and index maps.

3 The measurement records, handbooks of calculation and control survey, cross-section measurement achievements.

4 Section drawing, vertical-section and cross-section maps.

5 Technical summary reports.

11.9 Engineering Survey in Urban Water Projects

11.9.1 Engineering survey in urban water projects shall include the following contents: control survey, topographic survey, water engineering factor investigation, underground pipelines detection, etc., and shall be carried out according to the specific requirements of the measurement work at different design stages.

11.9.2 Engineering survey in urban water projects shall make full use of the existing urban measurement data and analyze the existing horizontal coordinate system, elevation system, control data and topography data, to make corresponding measurement plan.

11.9.3 Control survey shall meet the following stipulations:

1 The control survey should use the coordinates and elevation system of local urban survey.

2 Horizontal control network may be divided according to the existing local urban horizontal control network grade classification, and its accuracy indicators shall satisfy the corresponding stipulation specified in this standard.

3 In the stage of project proposal and feasibility study, existing control data shall be used to densify the partial control points to meet the need of topographic survey and other measurements. In the stage of preliminary design and construction, comprehensive network or densification control network may be laid out according to the project scale and the existing surveying control data.

4 Other technical requirements of horizontal control and vertical control survey shall meet the basic stipulations specified in Chapter 4 and Chapter 5.

11.9.4 Topographic survey shall meet the following stipulations:

1 The topographic engineering survey in urban water projects shall make full use of existing topographic map data. According to the needs of topographic map at different design stages, an economic and reasonable plan for topographic survey shall be made; as well as the topographic survey scope, scale, subdivision and supplementing survey method of topographic survey shall be determined.

2 Combining the engineering design stage with the city's existing basic survey data, suitable mapping scale shall be cho-

sen comprehensively according to the requirements of Table 11.9.4.

Table 11.9.4 The scale of topographic engineering survey in urban water projects

Project	Design stage	Mapping scale
Water diversion project	Project proposal, feasibility study	1 : 1000 - 1 : 10000
	Preliminary design, construction stage	1 : 500 - 1 : 1000 1 : 200 partially
Water environmental engineering	Project proposal, feasibility study	1 : 1000 - 1 : 2000
	Preliminary design, construction stage	1 : 500 - 1 : 1000 1 : 200 partially
Other water engineering		1 : 500 - 1 : 2000

3 For subdivision of topographic map, subdivision method of local urban topographic map, the subdivision of national norm map or the subdivision of free map may be used.

4 In project feasibility study, existing topographic data shall be made full use of in the supplementing survey, and the supplementing survey shall be carried out in area with scarce topographic data. In the stage of preliminary design and construction drawing design, the landform of entire construction area should be re-surveyed.

5 The elements of water engineering topographic survey shall be divided into the key terrain features and the general terrain features according to the contents of the project design. For key terrain features such as river drains, river bridges, roads drainage facilities, etc., survey method, accuracy indicators and representation method shall be specified in the measurement scheme. For general terrain details, the stipulated methods and accuracy in Chapter 6 may be used.

11.9.5 Water projects engineering elements investigation shall meet the following requirements:

    1 Investigations shall consider the water projects engineering project design and different design stages. The investigation mainly includes river outfall investigation, water pollution source investigation, urban demolition on-site investigation within the scope of engineering construction and other required investigation according to the design department.

    2 Specific investigations of river outfall include the horizontal coordinates and elevation of the outfalls, section dimension, sewage types, discharging forms, the filed photographs, discharge quantity, discharge quality, sounding, etc.

    3 Pollution source survey includes the related factories, living communities and breeding (planting) area pollutants investigations. Pollution source may be divided into point pollution source and non-point pollution source. Survey contents shall be determined according to the needs of the project. Survey method should be tracing relevant pollution sources back along the drainage pipeline. After the investigation, pollution source classification and distribution map shall be drawn.

    4 The quantity and area of the buildings, vegetation and other objects within the scope of permanent or temporary land requisition for water engineering construction shall be surveyed in compliance with the following requirements:

        1) Surveying methods include checking on-site, measuring on-site and measuring on the map, etc.

        2) The object classification and its standard shall comply with the land expropriation compensation files of the

nation and the local government.

3) For constructions survey, relevant data shall be measured on-site.

4) The survey of trees or other economic forests shall be determined according to specific situation of the design stage. The area of the trees covering land may be measured tree by tree or by natural parcels. When the area is measured by natural parcel, the number of trees may be calculated by the method of sampling survey.

5 Other element survey may be carried out according to the specific requirements of design units.

11.9.6 Survey and measurement of underground pipeline for water supply, rain water, sewage, gas, electricity, telecommunications, industry, etc., buried in the water project construction area shall meet the following stipulations:

1 The work contents of underground pipeline detection include the location, quantity, elevation, embedded depth, pipe diameter, material, etc., of each underground pipeline.

2 The accuracy of underground pipeline detection for water projects engineering shall meet the requirements specified in CJJ 61. The horizontal position deviation ΔS and depth deviation ΔH of hidden line points shall meet the following requirements:

1) Horizontal position tolerance shall be calculated according to Equation (11.9.6-1):

$$\Delta S \leqslant 0.1h \qquad (11.9.6-1)$$

2) Burial depth tolerance shall be calculated according to Equation (11.9.6-2):

$$\Delta H \leqslant 0.15h \qquad (11.9.6-2)$$

Where:

h = the buried depth of the underground pipeline center (cm). If $h \leqslant 100$cm, take $h = 100$cm in calculation.

3) For the detecting of deep-embedded pipelines or poor conductor pipelines, special detection technology design documents should be compiled according to engineering conditions in the field. And the pipeline detection units shall take various detection methods to find out its horizontal position and depth to provide predictive value, and shall excavate for verification before the construction or in construction process if necessary.

3 The mapping scale of underground pipeline survey should be 1 : 500 or 1 : 1000, and 1 : 2000 for area with sparse pipes.

4 The detecting depth of the underground pipelines may be determined according to requirements at different planning and design stages. At feasibility study stage, collecting existing information, field investigating and drawing sketches shall come first. At preliminary design stage, detailed detection and construction drawing shall be done. At design phase, the pipeline survey results at important parts shall be checked and the section map of intersection of pipeline and design line shall be drawn.

5 Before the project operation, original construction drawings, as-built drawings, present status drawings, management and maintenance data of underground pipeline in the surveying area shall be fully collected. After the pipeline detection at construction drawing stage, the related information shall be sent to the owner units of each line for advice.

6 Pipeline points should be set on feature points such as starting and ending points, turning point, branch points, diameter changing place, slope changing place, intersection point, material changing point, (into) outlet, center point of supporting facilities, etc. The interval of acquisition points at pipeline segment should be 10 - 30 cm on map; hidden line points shall be obviously signed.

7 The survey items and the choice standard of underground pipe lines should refer to the requirement of the client, and also to the density of pipelines, diameter size and importance specified in Table 11.9.6.

Table 11.9.6 The trade-off standard of underground pipeline survey

Classification of pipeline	Pipeline shall be detected
Water supply	Pipe diameter$\geqslant$50mm or pipe diameter$\geqslant$100mm
Drainage	Pipe diameter$\geqslant$200mm or box drain$\geqslant$400mm$\times$400mm
Gas	Pipe diameter$\geqslant$50mm or pipe diameter$\geqslant$75mm
Industry	All
Heating power	All
Electricity power	All
Telecommunication	All

8 For underground pipeline surveying, methods of field survey for clear line, probe method for hidden line points and excavation for difficult points may be used to determine the position of measured pipeline points and the property of surveyed pipeline.

9 Hidden metal pipes should be detected by direct method, clamp method or electromagnetic induction method, and hidden non-metallic pipes should be detected by photoelectric method or seismic wave method.

10 The permissible RMSE of pipeline points relative to neighboring control points shall be $\pm$5cm for plane position, and $\pm$2cm

for elevation measurement. The accuracy of underground pipeline survey shall satisfy that the permissible RMSE of interval of underground pipelines and adjacent constructions on the ground, road centerline or adjacent pipeline shall be ±0.6mm on the map.

   11   The excavation and investigation of underground pipelines shall be carried out safely. Excavation of electricity cables and gas pipes shall be done with professional staff. When investigation in well is implemented, the safety of workers shall be ensured and corresponding protective measures shall be taken.

11.9.7   As-built survey of underground pipelines shall comply with the following stipulations:

   1   The as-built survey of newly built underground pipelines shall be performed before earth up. If the survey cannot be carried out before earth up, measuring points of pipeline shall be set before earthing up and the location of measuring points shall be guided onto the ground accurately, and the description of station should be made.

   2   After the as-built of underground pipelines survey, final survey data and map of pipeline project shall be formed. The cross-section map of final pipeline survey shall be compiled if necessary.

   3   The data of as-built survey shall meet the requirements of local data loading.

11.9.8   After the completion of project, the following documents shall be collated and submitted:

   1   Technical design document.
   2   Field measurement records.
   3   Pipeline investigation records.
   4   Calculation data of measurement.
   5   Topographic map and comprehensive pipeline map.

    6 Control point result table, survey result table and pipeline detection result table.

    7 Technical summary report and inspection report.

11.10　Basic Control Survey in River Basin

11.10.1　Basic control survey in river basin includes basic horizontal control survey and basic vertical control survey.

11.10.2　Before the layout of river basic control network, the following information shall be collected:

    1 Existing national-grade plane and vertical control data within the river basin scope.

    2 The topographic maps and other related information within the river basin scope.

    3 The planning information of river basin.

11.10.3　The layout of river basin basic control network shall meet the following requirements:

    1 The control survey shall be laid out on the basis of the national geodetic control network.

    2 The control survey shall be implemented according to comprehensive design, hierarchical layout and integrated implement. The best design shall be selected according to the accuracy, reliability, cost, etc.

    3 The control network grade shall be determined according to the river basin scale, condition of the national geodetic control network, river basin planning needs, etc.

    4 The basic control network layout shall cover the main stream and the main tributaries, and meet the needs of water conservancy and hydropower engineering construction within the river basin.

11.10.4 The technology design of horizontal control survey shall comply with the following stipulations:

　　1 The horizontal control network shall be established by GNSS method or triangulateration method.

　　2 The horizontal control survey may be divided into B, C class or the second, third class, and the class shall be determined according to the stipulations specified in Table 11.10.4.

Table 11.10.4　Selection of horizontal control network class

Basin length	Method of measurement		
	Satellite positioning survey	Triangular control network survey	Traverse survey
Main stream length longer than 500km, and shorter than 1000km	class B, class C	Second class, third class	—
Main stream length shorter than 500km and tributary length not shorter than 100km	class C	Third class	Third class
Tributary length shorter than 100km	class C	Third class	Third class
Note: Control network of class B or class C for satellite positioning survey should be in compliance with standard GB/T 18314.			

　　3 If the main stream is longer than 1000km, the river basin basic horizontal control shall be designed specially.

　　4 In the survey area, national control points with higher grade than the control network shall be connecting surveyed. If the results are reliable, these national control points may be used as known points.

11.10.5 The elevation control survey design shall comply with the following stipulations:

1 The elevation control survey may be divided into the second class, third class and fourth class, which shall be determined according to the stipulations of the route length and the water surface slope specified in Table 11.10.5. The elevation control survey of each level should adopt leveling method. For the third and fourth class elevation control survey, trigonometric leveling with electromagnetic distance measurement may also be applied.

Table 11.10.5 Selection of elevation control network class

Route length (km)	Water surface slope		
	>1/5000	1/5000 – 1/16000	<1/16000
>200	Second class	Second class	Second class
80 – 200	Third class	Third class	Second class
<80	Fourth class	Third class	Second class

2 The original elevation control points near the leveling route shall be connecting surveyed.

3 The hydrological gauging station and observation basis points of gauging station within 1km away from the level route shall be measured with the same route lines, and for station and points within 5km, same class branch measurement shall be carried out.

11.10.6 Fixed surveying mark of river basin basic control points shall be buried. The mark structure and burying method shall be solid and suitable for permanent storage. For large town, convergence of tributary and planned engineering area, basic control points shall be set.

11.10.7 The observation, data processing, and other requirements of river basin basic control survey shall be implemented according to the relevant stipulations specified in Chapter 4 or Chapter 5.

11.10.8 For data processing, the transformation parameters between the previous vertical system (or reference) and the current vertical system in the river basin shall be calculated.

11.10.9 The river basin basic control network shall be regularly maintained after the establishment, and be resurveyed as needed. The re-survey interval shall be not more than 10 years.

11.10.10 For river basin basic control survey project, following documents shall be collated and submitted:

    1 Technical design document.

    2 Description of control points.

    3 The figure and leveling route map of the horizontal control points.

    4 Horizontal and elevation calculation data.

    5 Control point result table.

    6 Technical summary reports.

11.11 Monitoring of Regional Subsidence

11.11.1 Before the subsidence monitoring, the existing regional data including topographic maps at scale 1 : 10000 or 1 : 5000, control points, data of geology, hydrology, soil freezing degree and other information shall be collected. On site reconnaissance, unified plan and design also should be conducted.

11.11.2 The entire subsidence area shall be divided into different subsidence areas with different subsidence amount in accordance with the hydrology, geological conditions and the average annual settlement, as well as the factors of regional social and economic development situation. The division of the subsidence areas, spacing of subsidence monitoring points, re-survey cycle and selection of survey grade shall be implemented according to the stipulations in Table 11.11.2.

Table 11.11.2 Requirements of the subsidence point spacing, re-survey cycle and the survey grade

Average annual subsidence (cm)	Subsidence points spacing (m)	Re-survey cycle	Survey grade
1 – 3	1000 – 2000	3 – 5 years	First class
3 – 5	700 – 1000	1 – 3 years	First or second class
5 – 10	500 – 700	0.5 – 1 years	Second class
10 – 15	250 – 500	3 – 6 months	Second or third class
>15	<250	1 – 3 months	Third class

Notes: 1. In larger water areas, the subsidence point spacing is not limited to this table.
2. The third-class leveling should adopt with invar staff.

11.11.3 The regional subsidence monitoring network is composed of datum points, operating control points, subsidence monitoring points and leveling observation routes. The datum points shall be buried in stable area outside the subsidence region and shall be not less than 3 points or groups. The operating control points shall be laid out in stable area. Their number shall be determined by site conditions and monitoring requirements. The subsidence monitoring points shall cover the entire subsidence region and should be distributed in vertical – section, and the positions shall reflect the subsidence characteristics.

11.11.4 The datum points and the operating control points shall adopt rock marks, steel-pipe bench marks or bimetal bench marks. The subsidence monitoring points may use concrete bench marks, rock marks, deep buried steel-pipe bench marks or surveying marks setting up on the stable buildings according to the regional geology condition. The bench marks shall be surveyed after the stable period.

11.11.5 The subsidence monitoring network should adopt the current national height datum; if it is difficult to be connecting surveyed with the national datum points, an independent height system may be established.

11.11.6 In the early stage of monitoring, the whole subsidence monitoring network may be measured at the same accuracy. After getting some information about the subsidence area, the same accuracy or unequal accuracy measurement may be used according to the amount of average annual subsidence, the subsidence area and the actual requirements.

11.11.7 The instruments and technical requirements of subsidence network monitoring shall be implemented with relevant stipulations specified in GB/T 12897 and GB/T 12898.

11.11.8 In order to eliminate or weaken the influence caused by the non-unified subsidence between monitoring points or between leveling points in the subsidence process, the following measures shall be adopted in the subsidence observation:

1 The length of leveling observation ring or connecting traverse should be shortened. Two or more same type instruments may be used at the same time to shorten the observation time.

2 The route, season, instruments and scale plates of subsidence observation shall be relatively fixed.

3 The subsidence observation shall start from the regions with larger subsidence, and then gradually advance to regions with smaller subsidence. The measurement should be carried out simultaneously when high-class and low-class leveling route are observed in the same year (month).

4 In regions with large subsidence, the observation for a close loop shall be completed within a short time. If subsidence monitoring network nodes are observed cooperatively by several groups, connecting survey shall be done at the same time.

11.11.9 Before the adjustment computation for subsidence monitoring network, the stability of datum points and operating control points shall be tested.

11.11.10 The relevant data compilation and submission of regional surface subsidence monitoring shall meet the relevant stipulations in Section 5.6. And the following calculations shall be carried out and the following data shall be prepared:

1 Calculating the current subsidence, the accumulated subsidence and subsidence velocity or average annual subsidence of each monitoring point.

2 Calculating the current subsidence, the accumulated subsidence and subsidence velocity or average annual subsidence of each subsidence partition and the entire region.

3 Drawing yearly or monthly subsidence - time curve of abnormal subsidence points.

4 Drawing the isograms of subsidence according to current subsidence or subsidence velocity of the monitoring points; the isogram interval may be determined by subsidence or requirements.

5 Compiling the report of regional surface subsidence monitoring.

11.12 Survey for Engineering Construction Control Network

11.12.1 This section applies to the control network survey of medium or above water conservancy and hydropower project in the construction stage. The construction control network survey of small water conservancy and hydropower project may refer to this section.

11.12.2 The general requirement of horizontal control network shall conform to the following requirements:

　　1 The horizontal control network may be laid out to be the triangular network, GNSS survey network, traverse network or their combination types. The triangular networks and GNSS survey networks may be subdivided into the second, third and fourth class, the traverse networks may be subdivided into the third and fourth class. Every type and class may be used as the primary control network. The scopes of application are shown in Table 11.12.2.

Table 11.12.2　Class of primary horizontal control network for projects

Project scale	Primary class of horizontal control network	
	Concrete structure	Earth and rock structure
Large	Second, third class	Third, fourth class
Medium	Third, fourth class	Fourth class
Note: The concrete structure of water conservancy and hydropower project with special requirements may select first-class control network, which should be specially designed.		

　　2 The class of horizontal control network layout may be decided by engineering scale topographical condition and setting-out requirements. The first and second class is appropriate. The allowable root mean square error (RMSE) of the location of the first class control points shall be $\pm (5-7)$ mm for large water conservancy and hydropower project, and $\pm (7-10)$ mm for medium water conservancy and hydropower engineering. The allowable mean square error of the last class of horizontal control network point relative to the primary class control network point shall be ±10mm.

 3 The starting points of horizontal control network shall be placed near the dam axis or the main structure. For the horizontal control network of long headrace tunnel in water conservancy and hydropower project, the starting points may be set in dam area or plant area in the unified control system of primary control network, to ensure the relative rigor of each districts in the same control system.

 4 The primary control network should be an independent network. Mapping control points laid in the planning and design stage may be used as the initial data; if it is possible, these control points should be connecting surveyed with the adjacent national class control points, and the accuracy of connecting survey shall not be lower than that of the fourth class control network.

 5 The observational data of horizontal control network may skip the Gauss projection correction and direction correction; the side lengths shall be converted to the height surface selected by the project, and they may be calculated directly in plane rectangular coordinates system.

11.12.3 The reconnaissance, setting monument and surveying marks for horizontal control network shall meet the following requirements:

 1 The horizontal control network points shall be set in the place with good intervisibility, convenient transportation, stable foundation and long term preservation. The control network using GNSS shall also meet the relevant requirements for satellite signals receiving.

 2 The primary control points, which may be preserved for a long time and far away from construction area, shall be convenient for densification, and which are near construction area, shall be convenient for setting-out.

3 A concrete observation post with forced centering device shall be buried at the primary control network points and the control points on the main axis of the main buildings. When densifying network points, if it is inconvenient to bury a concrete observation post with forced centering device, the steel frame marks or ground marks may be buried.

4 The top surface of the forced centering device shall be buried horizontally; the non-flatness shall be less than 4′. The allowable deviation of forced centering for observation instrument and sighting equipment should be 1mm.

5 Striking protective device or mark around the control points shall be set according to the on-site conditions.

11.12.4 Horizontal control network survey shall meet the following requirements:

1 The main technical requirements for each class triangulateration network and trilateration network shall meet the stipulations specified in Table 11.12.4-1.

2 The main technical requirements for the third and fourth-class traverse network shall meet the stipulations specified in Table 11.12.4-2.

3 The main technical requirements for horizontal angle observation shall meet the stipulations specified in Table 11.12.4-3.

4 The main technical requirements for electro-optical distance measurement shall meet stipulations specified Table 11.12.4-4.

5 The main technical requirements for GNSS measurement shall meet stipulations specified in Table 11.12.4-5.

Table 11.12.4 – 1　Technical requirements for triangulateration network and trilateration network

Class	Side length (m)	RMSE of angle obs. (″)	Max. closing error of triangle (″)	Relative RMSE of mean side length	Nominal accuracy of distance measurement instrument (mm/km)	Side length	Number of round			
							Horizontal angle		Zenith distance	
							DJ1	DJ2	DJ1	DJ2
Second	500 – 1500	±1.0	±3.5	1/250000	±2	Two for direct and inverse	9	—	4	—
Third	300 – 1000	±1.8	±7.0	1/150000	±3	Two for direct and inverse	6	9	3	4
Fourth	200 – 800	±2.5	±9.0	1/100000	±5	Two for direct and inverse	4	6	2	3

Notes: 1. One round of electro-optical distance-measuring instrument is the process to sight once with 4 distance readings.
2. For a few special terrain conditions, it may be appropriate to relax the limit of side length. The relative RMSE of the weakest side may only be used as a reference.

Table 11.12.4-2 Technical requirements for connecting (close) traverse of electro-optical distance measurement

Class		Total length of close or connecting traverse (km)	Average side length (km)	RMSE of angle observation. (")	RMSE of distance measurement. (mm)	Relative closure error of total length	Closure error of azimuth (")	Nominal accuracy of distance measurement instrument (mm/km)	Side length	Number of round			
										Horizontal angle		Zenith distance	
										DJ1	DJ2	DJ1	DJ2
Third class		3.2	400	±1.8	±5	1/55000	±3.6$\sqrt{n}$	±3	Two for direct and inverse	6	9	3	4
		3.5	600		±5	1/60000							
		5.0	800		±2	1/70000							
Fourth class		1.8	300	±2.5	±7	1/35000	±5$\sqrt{n}$	±5	Two for direct and inverse	4	6	2	3
		3.0	500		±5	1/45000							
		3.5	700		±5	1/50000							

Note: The data in this table are calculated according to the requirement that the allowable RMSE of center point of straight connecting traverse shall be ±10mm.

Table 11.12.4-3 Technical requirements for the direction observation method of horizontal angle   Unit: (″)

Class	Type of instrument	Difference between two double readings of optical micrometer	Difference between readings of two sightings	Misclosure of round in half round	Comparative difference of 2C in one round	Comparative difference of every round in the same direction
Second, third, fourth class	DJ07, DJ1	1	4	6	9	6
Third, fourth class	DJ2	3	6	8	13	9

Note: If the difference of vertical angles between two observation directions is larger than ±3°, the 2C values may not be compared. The 2C value of each direction should only be compared with the same direction and adjacent round, and the difference shall still follow the stipulations in this table.

Table 11.12.4 – 4 Technical requirements for electro-optical distance measurement

Class	Nominal accuracy of distance measurement instrument (mm/km)	Tolerance of distance measurement			Meteorological data			
		Discrepancy of readings in one round (mm)	Discrepancy between rounds (mm)	Discrepancy of direct and reverse observations or optical section (mm)	Minimum reading of temperature (°C)	Minimum reading of atmospheric pressure (Pa)	Time interval of measurement	Data selection
Second class	±2	2	3		0.2	50	Starting and ending time of every side observation	Mean value of two ends of every side
Third class	±3	3	5	$2(a+bD)$	0.2	50	Starting and ending time of every side observation	Mean value of two ends of every side
Fourth class	±5	5	7		1	100	One time for every side	Observation value of the station

Notes: 1. One round of electro-optical distance measurement is the process including one sighting and 4 readings of distance.
2. The discrepancy of direct and inverse may be compared after the slope distances were converted to the same height surface.

Table 11.12.4-5　Main technical requirements for different classes GNSS surveying control network

Class	Side length (m)	Nominal accuracy of instrument		Relative RMSE of mean side length	Side numbers of close loop or connecting traverse
		a (mm)	b (mm/km)		
Second class	500~2000	≤5	≤1	1/250000	≤6
Third class	300~1500	≤5	≤2	1/150000	≤8
Fourth class	200~1000	≤10	≤2	1/100000	≤10

Notes: 1. The mode of static operation shall be adopted in GNSS control network survey, and the dual-frequency receivers shall be used if the second or third class horizontal control survey is implemented.
2. The basic requirements for static operation of different class GNSS survey shall comply with Section 2 of Chapter 4.

6 The record of direction (angle) observation, electro-optical distance field observation should adopt the data recorder or data terminals.

11.12.5 The general stipulations of vertical control network shall in compliance with the following requirements:

1 The vertical control survey may be subdivided into the second, third and fourth class. Each class may be used as the primary control network. The control network class may be determined according to the project scale, the scope and accuracy of setting out. Their working ranges are shown in Table 11.12.5.

Table 11.12.5　Class selection of primary vertical control network for different projects

Project type	Class of primary vertical control network	
	Concrete structure	Earth and rock structure
Large	Second class	Third class
Medium	Third class	Fourth class

2 The vertical control network may be established by one measurement or a combination of several measurement types such as leveling, cross-river leveling, trigonometric leveling and transferred vertically measurement using precise steel rule, according to the landform and traffic conditions of survey area.

   3 The accuracy requirement of vertical control network: the permissible RMSE of height of the lowest vertical control network point relative to the primary control network point shall be ± 10mm for concrete structure, ± 20mm for earth and rock structure, and for hydraulic tunnel it should conform to the relevant stipulations of SL 52.

   4 The primary vertical control network and the densified vertical control network should be laid into the close loop, the connecting traverse or the network with junction points. Spur leveling line should not be laid out.

   5 The primary vertical network should be connecting surveyed with the neighbor nations' bench marks, and the connecting survey accuracy shall not be lower than that of the fourth-class leveling.

11.12.6 Reconnaissance and monumentation of vertical control network shall meet the following requirements:

   1 The vertical control points shall be laid at the place that will not be affected by construction and be convenient for long time protect and use. For the dam, the vertical control points should be evenly distributed on the left and right bank of the upstream and downstream of the dam axis without the influence of flood. For hydraulic tunnel, the vertical control points shall be set at the inlet and outlet of the tunnel, near the entrance of branch tunnel or shaft. Their elevation should be close to the bottom elevation of the excavated tunnel.

2 At least 2 vertical control points should be laid near each individual project portion or each construction tunnel inlet.

3 For the vertical control points, bedrock marks or concrete bench monuments may be embedded. And some horizontal control points can be used as vertical control points.

11.12.7 The observation of vertical control network shall meet the following stipulations:

1 The main technical requirements for the second, third and fourth class leveling, cross-river leveling and the fourth-class trigonometric measurement shall comply with the stipulation in Chapter 5.

2 When using precise steel ruler to transfer height vertically, it shall be ensured that the tension of ruler is consistent with the tension in verification. The upper and lower ends shall be simultaneously observed and the temperature shall be recorded.

3 Records of field observation should adopt data recorder or data terminals.

11.12.8 The checking calculation and adjusting computation for the construction control network survey result shall comply with the following stipulation:

1 Before the adjustment computation, the field note and the initial data for adjustment computation shall be checked completely.

2 After finishing every observation of construction control network, each tolerance shall be checked according to the type of the control network survey.

3 Each class of the construction control network shall be adjusted rigorously.

11.12.9 After the construction control network has been established, the maintenance and management shall be conducted carefully, which includes the following two aspects:

 1 Resurvey the control network to identify and correct possible displacement in time.

 2 Expand and densify the network points in time with the progress of the project, to meet the setting-out requirements.

11.12.10 After the construction control network is established, resurvey shall be conducted under the following situations:

 1 One year later after the control network is established.

 2 When the excavation is finished, and during the concrete project and the installation of metal structure, mechanical and electrical instrument.

 3 If some control network points are collided, or cracks or new project are found around it.

 4 If there is a felt earthquake.

 5 If the control point is used as initial data to lay out local control network.

 6 The control points located in high slope area or near excavation area should be resurveyed more frequently.

11.12.11 Resurvey of construction control network shall comply with the following stipulation:

 1 Resurvey of the construction control network may adopt the whole network or the local net point according to the specific conditions. The resurvey accuracy should not be less than the accuracy of control network establishment.

 2 The fixed point or quasi-stable point of resurvey may be selected according to their reliability and positions in the network. Several more fixed points or quasi-stable points may be selected in the adjustment computation of resurvey network. Based on the size and distribution of correction, phase out displacement points or increase stable points, and correctly identify the displacement of network points.

11.12.12 The construction control network survey project shall collate and submit the following data:
    1 Technical design document.
    2 Construction control network map.
    3 Description of control points.
    4 Inspection data of measuring instruments and equipment.
    5 Original filed observation handbooks and project estimation data.
    6 Results of adjustment calculation.
    7 Technical summary reports.

11.13 Survey for Engineering Deformation Monitoring Network

11.13.1 The survey for monitoring deformation network shall be established for the deformation monitoring of the hydraulic structure, rock mass and high-slope near the construction area, unstable object at the bank and slope, and the high-cut slope of construction of reservoir resettlement towns, along with the fixed point monitoring of crustal deformation in the reservoir area.

11.13.2 The classification and accuracy requirements for deformation monitoring shall meet the stipulations specified in Table 11.13.2.

11.13.3 The general requirement of deformation monitoring network shall comply with the following requirements:
    1 The establishment of deformation monitoring network shall set the datum points; if the long distance between the datum points and survey object leads to insufficient deformation measurement accuracy or the operation is inconvenient, the op-

erating control points shall be set.

Table 11.13.2 Classification and accuracy requirements for deformation monitoring

Class	Vertical displacement monitoring — Allowable RMSE of displacement value (mm)	Horizontal displacement monitoring — Allowable RMSE of displacement value (mm)	Application range
First class	±1.0	±2.0	Large concrete hydraulic structure, large rocky high slope, crustal deformation in large reservoir-dam area, etc
Second class	±2.0	±3.0	Large and medium hydraulic structure sensitive to deformation, large high slope and high cut slope, collapse and landslide of large bank slope, etc
Third class	±3.0	±5.0	Earth and rockfill dam engineering project, collapse landslide of general side slope and bank slope

Note: The RMSE of deformation monitoring displacement usually is $\sqrt{2}$ times as the RMSE of single measurement.

2 The deformation monitoring network consisting of datum points, operating control points and deformation monitoring points, may be laid out into two layers, the monitoring reference network and the monitoring operation network, and shall meet the following rules stipulations:

1) The monitoring reference network consists of datum

points and operating control points to check the stability of datum points and operating control points.

2) The monitoring operation network consists of some datum points, operating control points and deformation monitoring points to observe displacement of monitoring points.

3) The layout of monitoring reference network and monitoring operation network may be laid and ungraded, and they should be observed with the same accuracy.

4) The design of monitoring network shall suit the local conditions. The method and class of observation shall be determined according to requirements of the construction scale, monitoring accuracy, as well as the reliability and sensitivity of net, etc.

3 The horizontal displacement monitoring network may be observed by the methods of triangulation, trilateration, combination of triangulation and trilateration, and GNSS survey, etc.

4 The vertical displacement monitoring network should be observed by the leveling method and shall be laid into a connecting traverse or a close loop. And the route may be supplemented with vertical height transfer, trigonometric leveling, etc. , if needed.

11. 13. 4 The selection and burying of the monitoring datum points and operating control points shall conform to the following requirements:

1 The datum points of horizontal or vertical displacement monitoring shall not be less than 3.

2 The datum point shall be set on solid bedrock or soil base outside the deformation area. If the base of datum point does not meet the stable requirements, foundation reinforcement shall be carried out. The operating control point shall be set in a relatively stable region near the observation area. It shall be noted that the operating control point location should form an observation network with monitoring points, which is good for observation accuracy. Besides, the location shall be convenient for observation.

3 A concrete post with forced centering device shall be established at the datum point and operating control point of horizontal displacement monitoring.

4 Bedrock monuments and concrete monuments are the best options for the datum points of vertical displacement monitoring. Bimetal bench mark or deep buried steel-pile bench mark shall be buried in deformation area, if the condition is limited. If possible, adit mark may be adopted.

5 For the operating control point of vertical displacement monitoring, rock monuments, concrete monuments, bimetal bench mark, steel-pile bench mark may be used; the bench mark may also be set on the wall of a stable building.

11.13.5 The observation of horizontal displacement monitoring network shall conform to the following requirements:

1 The main technical requirements for horizontal displacement monitoring network shall meet the stipulation in Table 11.13.5-1.

2 The direction observation method should be used for the observation of horizontal angle. The first class observation shall be implemented according to relevant stipulations in SDJ 336. The second class and third class observation shall be implemented according to the stipulations in Table 11.12.4-3.

3 The side length of monitoring network may be measured by electro-optical distance measurement. Its main technical requirements should meet the stipulations in Table 11.13.5-2.

Table 11.13.5-1 Main technical requirements for horizontal displacement monitoring network

Class	RMSE of location (mm)	Mean side length (m)	RMSE of distance observation (mm)	RMSE of observation side	RMSE of angle observation (")	Number of round of horizontal angle	
						DJ1	DJ2
First	±1.5	300 - 700	$1 + 1 \times 10^{-6} D$	$\leqslant 1/300000$	±0.7	12	—
Second	±2.0	400 - 1000	$1 + 1 \times 10^{-6} D$	$\leqslant 1/200000$	±1.0	9	—
Third	±3.5	800 - 1800	$1 + 1 \times 10^{-6} D$	$\leqslant 1/100000$	±1.8	6	9
		300 - 800	$2 + 2 \times 10^{-6} D$				

Notes: 1. D is the length of measurement side, km.
2. The RMSE of location is the RMSE of network point relative to the adjacent datum point or reference point.

4 The GNSS monitoring network shall be specially designed. The RMSE of location shall meet the requirements in Table 11.13.5-1.

11.13.6 The observation of vertical displacement monitoring network shall be implemented according to the following requirements:

1 The first and second class observation of vertical displacement monitoring network shall use direct leveling and conform to relevant stipulations in GB/T 12897.

2 The third class observation of vertical displacement monitoring network shall conform to the stipulations in GB/T 12898, but the instrument shall adopt DS05 or DS1 level and its leveling staff.

Table 11.13.5-2 Main technical requirements for electro-optical distance measurement

Class	Accuracy of distance measurement instrument (mm/km)	Number of round at each side		Discrepancy of readings in one round (mm)	Discrepancy of round in a single range (mm)	Minimum readings of meteorological data		Discrepancy of direct and reversed observations (mm)
		Direct observation	Reverse observation			temperature (°C)	pressure (Pa)	
First	$\leqslant 2$	3	3	2.0	2.0	0.2	50	$\leqslant 2(a+bD)$
Second	$\leqslant 2$	3	3	2.0	3.0			
	$\leqslant 2$	2	2	3.0	4.0			
Third	$\leqslant 4$	3	3	4.0	6.0			

Notes: 1. One round is the process including one sighting and 4 readings.
2. According to the specific circumstances, side measurement may adopt observations at different times instead of direct and reversed observation.
3. The meteorological data shall be observed and recorded at each end of side measurement. The average meteorological value of two end points shall be used to the meteorological correction calculation.
4. After the meteorological correction and the correction of additive and multiplicative constant, the slope distance was able to convert to the horizontal distance.

3 For a few monitoring network points, at which are difficult to be measured by leveling, trigonometric leveling may be adopted. Vertical angulation shall use two instruments with accuracy not lower than that of DJ1 to reciprocal observe 12 sets at the same time. The 12 sets shall be divided into two periods and use direct and reverse observation 6 times respectively. The index difference of vertical angle between rounds shall not exceed 7″, mutual difference of vertical angle shall not exceed 5″.

4 The other requirements for observation of vertical displacement monitoring network shall meet the relevant stipulation in Chapter 5.

11.13.7 The first survey and resurvey of deformation monitoring network shall meet the following stipulations:

1 The deformation monitoring network shall be built with the progress of the project. The first observation shall be completed before the dam first - impoundment, and the first value may be used as the datum value. The first observation shall be observed at least twice continuously and separately. After the observation is qualified, the average value may be taken as the first observation value.

2 The observation cycle of deformation monitoring network shall be determined by a lot of factors, such as the deformation characteristics, deformation rate, observation accuracy, geological conditions of the project and operation management, etc. During the monitoring, the observation cycle should be adjusted properly according to the change of deformation.

3 The deformation monitoring reference network shall be measured once a year. The deformation monitoring operation network shall be measured every 1 - 3 months. In special cases such as reservoir impoundment, felt earthquake or accelerated deformation, etc., measurement frequency shall be increased.

11.13.8 The observation of deformation monitoring network for each period shall meet the following requirements:

1 Finish in a short time.

2 Use the same figure (observation route) and observation method.

3 Use the same type of instrument and equipment.

4 Same observers.

5 Record relevant construction and environmental factors such as load, temperature, precipitation, water level, etc.

6 Process data under the same reference.

11.13.9 The data processing and stability analysis of deformation monitoring network shall be implemented according to the following requirements:

1 The original records of deformation monitoring network shall be arranged and checked in time.

2 The correction calculation and verifying calculation of observation data and data processing shall comply with the stipulations in Chapter 4 and Chapter 5. If there are no clear definitions in this code, the relevant national or industry stand should be followed.

3 The adjustment computation of monitoring reference network and monitoring operation network should be successively calculated.

4 The reference point of monitoring reference network adjustment shall be the point or point group which passed the stability test. The stability test of datum point position may adopt the following methods:

　　1) Comparing the observations: compare the difference

(change) between the first survey result and resurvey result of two points. If the difference is less than the maximum measurement error (twice of the RMSE), the two points may be considered to be relatively stable or have no significant change.

2) The survey adjustment of least squares method may be used for test. If Δ meets the requirements of the Equation (11.13.9), the two points are considered as to be stable.

$$\Delta < 2\mu \sqrt{2Q} \qquad (11.13.9)$$

Where:

Δ = the difference between adjusted values of resurvey and the first survey, mm;

μ = RMSE with weight unit, it may be the mean value of RMSE with weight unit in two periods;

Q = weight coefficient of deformation of observation point.

3) Mathematical statistics.

4) The combination of the three methods above.

11.13.10 The following data shall be sorted and submitted for the measurement project of engineering deformation monitoring network:

    1 Technical design document.

    2 The maps of deformation monitoring network.

    3 The description of station and mark stones burying maps of mark for deformation monitoring network.

    4 Inspection data of measuring instruments and equipment.

    5 The filed note of original observation data and the data of estimate.

    6 Adjustment calculation results.

    7 The technical summary reports.

11.14 Slope and Reservoir Bank Deformation Monitoring

11.14.1 The unstable region (landslide, collapse, immersion, etc.) of slope and bank, which threatens the safety of water conservancy and hydropower project as well as the safety of people's lives and property, shall be monitored.

11.14.2 The general requirement of the deformation monitoring of unstable region of slope and bank shall be executed according to the following requirements:

   1 The main content of deformation monitoring shall include perambulation inspection, horizontal displacement monitoring, vertical displacement monitoring, and crack monitoring etc., and collecting the possible causing reason documents.

   2 The accuracy of deformation monitoring shall meet the stipulation specified in Table 11.14.2.

Table 11.14.2 Deformation monitoring accuracy of unstable slope and bank   Unit: mm

Region	Items	Allowable RMSE of displacement
Slope	Horizontal displacement	±3 (rock), ±5 (soil)
	Vertical displacement	±3 (rock), ±5 (soil)
	Crack	±1
Unstable region of bank (landslide, collapse, immersion, etc.)	Horizontal displacement	±5
	Vertical displacement	±3
	Crack	±1

Notes: 1. The deformation monitoring accuracy of special circumstances such as extra-large slopes or landslides, etc., may be determined in the design according to the actual situation.
2. The monitoring accuracy of the rock mass and high slopes at the near dam area shall refer to SDJ 336.

3 Before the deformation monitoring, relevant hydrogeological data, geotechnical engineering data and design drawings shall be collected. In addition, the deformation monitoring plan shall be designed according to the factors such as geotechnical condition, project type and construction method etc.

4 During the deformation monitoring, the instruments and equipment shall be checked, adjusted and calibrated regularly, and records should be made.

5 The deformation monitoring shall comply with the following stipulations:

    1) The datum points of deformation monitoring shall be set in stable regions and shall not be less than 3.

    2) The monitoring points shall be set at the places which may reflect deformation characteristics of the soil and rock mass.

    3) The monitoring equipment shall be protected properly.

    4) The equipments used in deformation monitoring shall meet the accuracy requirements specified in the Table 11.14.2. The same project shall use the same equipments.

    5) Relevant monitoring projects should be conducted at the same time in a favorable period.

6 Each reference value shall be observed independently and continuously at least twice. When the observation is qualified, the average value shall be used as reference value.

7 The monitoring data shall be compiled and analyzed in time.

11.14.3 The design of deformation monitoring shall meet the following requirements:

    1 The monitoring points shall be set according to cross-sections. According to the landform and geological survey data, the vertical-section with landform and geological characteristics shall be chosen as monitoring vertical-section.

2 The surface vertical displacement monitoring points are recommended to set at the same position with horizontal displacement monitoring points. The deep monitoring points observed by inclinometers and multi-point displacement meter shall be arranged with the surface horizontal and vertical displacement monitoring points.

3 The surface horizontal displacement may be monitored by the methods such as intersection, triangulateration network, collimating line method, polar coordinate difference, GNSS measurement, laser scanning measurement, etc. ; the deep displacement maybe monitored with inclinometers, multi-point displacement meter, etc.

4 The vertical displacement may be monitored by the methods such as leveling, EDM trigonometric leveling, polar coordinate difference, GNSS measurement, etc.

5 The design, burying, observation and data processing of the datum points and operating control points of horizontal and vertical displacement monitoring shall follow the stipulations in Section 11.13.

6 Alignment observations should adopt minor angle method and moving-target method; the stadia length for rock slope should not exceed 300m and for soil slope should not exceed 800m.

7 Intersection method for horizontal displacement monitoring may be angle (direction) intersection, linear intersection, linear angular intersection, etc. The angle of angle (direction) intersection method should be between 60° to 120°. The angle of

linear intersection method should be between 30° to 150°.

8 The arrangement of observation station point and back sight point for polar coordinate difference method should meet the followings stipulations:

1) Observation station point and back sight point should be arranged near the stable region outside the slope or landslide region.

2) The elevation difference between observation station point and back sight point should not exceed 100m. The side length should not exceed 1000m. If possible, one back sight point may be set at each side of the monitoring area.

9 The GNSS method for deformation monitoring shall adopt the static relative positioning observation to conduct regular monitor or real-time monitor. The design of GNSS survey network shall meet the following stipulations:

1) The arrangement of GNSS survey network should adopt network connection, edge connection or hybrid connection.

2) The edges of independent close loops should not be more than 4; the independent baselines connected with monitoring points in the network shall not be less than 2.

3) The GNSS monitoring network shall be connecting surveyed with 2 or more than 2 stable datum points or operating control points.

4) If possible, some high precision side length may be measured additionally.

10 Various monitoring methods shall be designed according to the actual situation, the special requirements and specific conditions of monitoring objects. Its accuracy shall be estimated by least squares method and also make sure that the RMSE of displacement monitoring points not exceed the value specified in Table 11.14.2.

11 When GNSS method is adopted to monitor vertical displacement, the geodetic height may be used to calculate the vertical displacement. When use GNSS to fit height, the monitoring network points which are evenly distributed and have the high accuracy level height results shall not be less than 4.

12 The incline hole and multipoint displacement meter shall go through potential slip or disturbed belt (layer) to relatively stable region. The drilling and installation shall meet the stipulations in SDJ 336.

11.14.4 The observation of deformation monitoring points shall meet the following requirements:

1 The technical requirements of intersection method and triangulateration network method should be implemented according to the stipulations in Article 11.12.4.

2 The measurements by collimating line method shall adopt the angle measuring instrument with accuracy not lower than 1″. For the moving target method, at least 2 rounds shall be conducted, and their mutual deviation shall be less than 3mm. For the minor angle method, at least 4 rounds should be conducted, and the difference of minor angle of each round shall not exceed 2″.

3 The polar coordinates difference measurement shall meet the following stipulations:

1) For the polar coordinates difference method, transits of DJ1 or above and class I rangefinder should be used for survey. The main technical requirements shall meet the stipulations set out in Table 11.14.4.

Table 11.14.4 Main technical requirements of polar coordinates difference observation

Class	Nominal accuracy of instrument		Tolerance of displacement RMSE (mm)	Maximum length (m)	Observation set of horizontal angle	Observation set of length	Observation set of vertical angle
	Accuracy of distance measurement (mm/km)	Accuracy of angle measurement (″)					
First class	±2	±0.5	±3	500	12	4	6
			±5	1000			
Second class	±2	±1	±3	300	9	2	4
			±5	500			

2) The measurement error of instrument height and target height shall be less than ±1mm, and the estimate value of reading is to 0.1mm.

3) When using the polar coordinate difference method for horizontal or vertical displacement monitoring, the observation for one station or one period shall be conducted in stable meteorological conditions and completed as soon as possible.

4 The main observation technical requirements of GNSS measurement shall meet the relevant stipulations in Section 4.2.

5 When using trigonometric leveling with electromagnetic distance measurement to monitor the vertical displacement, two or three directions intersection measurement should be used. The technical requirements of observation shall meet the relevant stipulations in Section 5.4.

6 The monitoring period of unstable bank should be once a month. It may be appropriately adjusted according to drought, rainy season or deformation rate, etc.

7 During the construction period, the slope monitoring cycle should be once a week; during the operation period, the slope monitoring cycle may be once a season. It may be adjusted according to deformation rate, etc.

11.14.5 The following data shall be collated and submitted for the deformation monitoring project of slope and bank:

1 Technical design document.

2 Observation result and displacement.

3 Distributing map of monitoring points, layout map of datum network, directional map of observation and leveling route map.

4 Graph of horizontal displacement value and time, and subsidence chart.

5 Relative graphs of excavation against horizontal displacement, relative graphs of excavation against subsidence, graphs of displacement (horizontal and vertical) velocity against time.

6 Graph of deep displacement.

7 Monitoring report.

12 GIS Development

12.1 General Requirement

12.1.1 According to building methods, building principles and characteristics of GIS, GIS development can be divided into the following stages: requirements analysis, overall design, detailed design, software coding and testing, system operation and maintenance, etc. The main work content of each stage shall be performed on the basis of the following requirements:

1 During the stage of requirements analysis, the target content of GIS development shall be determined, and the system logic model shall be confirmed through close cooperation and full exchange between system analysts and users. Make clear the functions, performance requirements, design constraints as well as check and acceptance standard of the system, finally write the requirements analysis report.

2 During the stage of overall design, the overall structure of GIS shall be determined according to the result of system requirements analysis. The system analysts and designers shall have deep understanding of the system demands, and then implement the system design of the architecture, hardware and software configuration, function modules, database, system interface, reliability, etc. Finally, the overall design report shall be worked out.

3 During the stage of detailed design, programmers shall implement the system design including the detailed function design of each module, interface design as well as design of input and output, and write the detailed design report according to the

overall design requirements.

    4 During the stage of software coding and testing, the task is completing the software coding and testing, and the preparation of user manuals, test reports and other documents.

    5 During the stage of system operation and maintenance, the system functional structure shall be modified based on feedback from users to improve the quality of the system, making it more stable, and better in line with actual demand.

12.1.2 Prototyping is appropriate when system designing. The life-cycle method or module structure method may also be used for system designing when requirements analysis is clear.

12.1.3 GIS design shall follow the following principles:

    1 The user-oriented principles shall meet the following requirements:

        1) Practicability: the system design shall consider the technical methods and achieving means, as well as the storage, maintenance and update of big data.

        2) Extendibility: data codes, system function, data structure, application area, configuration of hardware and software shall be extendible.

        3) Security: the system shall have the necessary security and confidentiality measures, and a strong ability of dealing with computer crime and the virus prevention.

        4) Advancement: the hardware and software including operating system, database management system and GIS software shall be advanced and have mature technology.

    2 Standardizing principles shall meet the following requirements:

1) System content, data classification and codes, data accuracy, and operating procedures should adopt relevant national standards, professional standards and provincial standards.
2) The provisional regulations may be supplemented if the standardized contents are needed but there are no provision in national standards, professional standards and provincial standards.

3　The optimization principle of cost-effectiveness shall meet the following requirements:

1) Data accuracy shall meet the application requirements.
2) Choose the program with the best cost performance.
3) Arrange the work sequence reasonably.
4) First make experiments, and then implement in large sale.

12.2　Demands Analysis

12.2.1　Demands analysis shall investigate users' business processes, tasks, organization form and the existing systems, make clear users' specific business requirements, functional requirements, performance requirements and other requirements of GIS, as well as system boundary and interface details with other systems. Finally, provide a user requirements analysis report.

12.2.2　Demands analysis shall include the following contents:

1　System requirements analysis shall meet the following requirements:

1) Status description and business requirement: make clear the users' business status and specific business requirements.

2) Function requirement: make clear functions that the system shall realize, including data input, data query, statistical analysis of data, data output, etc.
3) Performance requirement: make clear the system performance indicators, including constraints for response time, requirements for data accuracy, restrictions for storage capacity, etc.
4) Other requirements: make clear the requirements of system security and reliability, and the relations with other information systems that users are using.

2 Data requirements analysis shall determine the data necessary for realizing the system function and the data type the system will process. Carry out the investigation of data requirement; determine scale of spatial data, data size and increment, output content, method and format of data, as well as interface of spatial data and non-spatial data.

3 The requirements analysis of system development shall make clear person, software, hardware, schedule, user interface, and system environment which system development needs.

12.2.3 Requirements analysis shall meet the following requirements:

1 Demands investigation shall meet the following requirements:

1) Investigate the users' organization, daily business, status and improvement requirements of application systems, etc.
2) Investigate the existed GIS. Define the functions and business GIS may realize, as well as data needed for realizing these functions.
3) Investigate the system software, hardware, network

conditions and other resources that users are using.

2 The methods such as a face-to-face talk, telephone interview, questionnaires, information query may be used for acquiring users' requirements.

3 The process of requirements analysis shall meet the following requirements:

1) Investigation preparation: before the requirement investigation, make an investigation plan, determine the content and means for investigation, and then divide the tasks.

2) Requirement investigation: conduct the user requirement investigation according to the investigation content.

3) Write requirements analysis report: according to the results of the requirement investigation, determine the data flow of business processes and data structures which GIS will process, refine the system functions, find out the relationship between the system elements, interface characteristics and system boundaries, and then based on the requirement of function, performance and operating environment, exclude the irrational part of requirement investigation to complete the requirements analysis report. The requirements analysis report shall be written according to the requirement specified in Table F.0.1 of Annex F.

4) Evaluating the requirements analysis report: evaluate the correctness, completeness and description clarity of the requirements determined in the requirements analysis report in order that the requirements and description content specified in the requirements analysis

report is practicable. After the requirements analysis report is evaluated and accepted by the user and developer, it can be regarded as the basis for system development design and system check and acceptance during the system construction.

12.3 Overall Design

12.3.1 On the basis of the overall objectives and system scale determined in the demands analysis report, the system overall structure shall be confirmed, and the function modules shall be divided, the relationship between modules and the configuration of systems hardware and software shall be determined, the data structure of system shall be designed, and the technical specifications and standards of the system shall be stipulated.

12.3.2 GIS overall design shall include the following contents:

    1 The design of overall structure shall meet the following requirements:

        1) Overall structure: divide the system into different subsystems according to the specific requirements, each subsystem contains different functional modules. The system shall have data input, data query, spatial analysis, data output, as well as other modules.

        2) Function design: design the basic function of the system modules. Data input module shall have the functions including graphic input, attribute data input and data import, etc. Data query module shall have the functions including retrieval by spatial extent, attributive search by graph, graphic search by attributes, etc. Spatial analyst module shall have overlay

analysis, buffer analysis, etc. Data output module shall have vector plotting, raster plotting, report output, data export, etc.

2 The configuration of hardware and software shall meet the following requirements:

1) Hardware platform: hardware platform shall include computers, input devices, output devices, data storage devices, data backup devices, network and uninterruptible power supply, etc. Different types of hardware shall be marked with the model, quantity, performance indicators, technical advantages, special conventions, etc.

2) Software platform: software platform shall include operating system, GIS basic softwares and network softwares. It is necessary to explain the technical characteristics of different softwares and differentia from similar products at home and abroad, clearly explain the superiority and mark out name, manufacturer, version number, and technical requirements of these selected softwares.

3) Network architecture: network architecture design shall include network design principles, technical requirements, product selection, topological structure, transmission medium, interface, communication protocol, etc. It is necessary to explain network security technology to be used.

3 Database design shall meet the following requirements:

1) Data content: it is necessary to give a full description of all kinds of data contained in the database, including spatial data, attribute data, metadata, etc.

2) Selection of database management system: it is necessary to select the database management system based on system functional requirements and the technical requirements of GIS basic softwares, and demonstrate if the database management system meets the requirements of the system.

3) Logical design of database: make a description of database naming convention, classification and codes of data, principles for layering, as well as naming conventions of layer and file.

4) Selection of spatial data model: choose the appropriate spatial data model and storage format according to the user needs.

5) Data security design: including data backup and security, user management design, etc.

6) Compile database design.

4 Interface design shall meet the following requirements:

1) User interface: including screens, menus, prompt message design, etc. User interface shall be unified in style, simple and targeted.

2) Internal interface: design the interface among the modules of the system.

3) Hardware/software interface: design the hardware/software interface mode of the system.

5 Reliability process shall meet the following requirements:

1) Error message: describe the possible error or fault conditions during the system operation, including the mode, meaning and handling methods for the system output information.

2) Remedy: design and describe the remedy and solution for the problems that appear during the system operation.

12.3.3 The contents of overall design and database design shall comply with the requirements specified in Table F. 0. 2 and Table F. 0. 3 of Annex F. It is necessary to argue the report in respect of advancement, integrity, reliability, extendibility, and reasonableness of design.

12.4 Detailed Design

12.4.1 During the stage of detailed design, the subsystems and modules divided in the overall design shall be further refined. According to cohesion, coupling degree, function integrity and modifiability further divide the modules into ones with independent functions and proper scales. And each module shall be high cohesion and low coupling. Carry out module design, draw the constitutional diagram of the module structure, and describe the content and function of each module in detail.

12.4.2 GIS detailed design shall meet the following requirements:

1 Module design shall make program description of each module, including algorithms and program flow, input and output item, external interface, etc.

2 GIS interface shall be easy to learn, flexible, and has a description about interface layout, color scheme, menu form, menu layout, and dialogue operating type, etc.

3 Input and output design shall have an explicit stipulation about the content, style, type, format, media and precision of input and output on the base of overall design.

12.4.3 The contents of detailed design shall comply with the requirements specified in Table F. 0. 4 of Annex F. Detail design can be argued alone or argued with overall design.

12. 5 Software Coding and Testing

12. 5. 1 Software coding of GIS shall meet the following requirements:

    1 Before coding, compile the coding standard that programmer shall comply with according to the characteristics of program language.

    2 Coding shall follow the following special requirements:

        1) The system involves the processing, display and read-write of massive data, which has higher requirements for system hardware and network equipment. Moreover, algorithm and program flow of the system will greatly influence the running efficiency. Therefore, when writing programs system configuration shall be considered efficient algorithms and data structures should be used.

        2) Because contents which system processed are more, in order to ensure readability and stability of the program, it shall have a good design style. That is to say, the source program shall be merged into documents, the data specification shall be reasonable, the structure shall be distinct and normative, module interface shall be clear, etc.

        3) In the process of system operation, it may appear many irregularities and even illegal operations. In order to avoid the system getting into a collapsed state, the program shall have a strong fault tolerance.

12. 5. 2 After GIS software coding, the system testing shall be carried out according to the following requirements:

1　The content of system testing shall meet the following stipulations:

 1) Unit testing or module testing: test the correctness of the smallest unit or program modules in software design.

 2) Integration test: on the basis of unit test, assemble all modules into a system and make a joint test according to the requirements in the overall design. Discover and rule out problems appearing in the module combination. Finally, a system which is in line with the overall design requirements will be constructed.

 3) Checking and acceptance testing: test the function and performance of the system and judge it meets the user requirements in the requirements analysis report. Users shall take participate in and design the acceptance test case. Input the test data through user interface, and analyze the output results. In addition to the functional testing, it shall also test the portability, compatibility, maintainability, fault tolerance performance and so on. At the same time, it is necessary to test the integrated environment of the whole system including the software, hardware, networks and other equipment to check if there are some parts which do not match the overall design, as well as find errors in the system analysis and design.

2　In order to test different functions, testing cases shall meet various requirements, and some erroneous data shall be used. And the relationship between the data shall comply with procedural requirements.

3 The steps of testing methods shall meet the following requirements:

1) Testing plan: make a complete testing plan before testing, including testing the functions of the system, testing contents, scheduling of each test, resource requirements, testing data, testing tools, selection of testing cases, mode and process of controlling tests, assemble mode of system, evaluation criterion, etc.

2) Testing organization: when organizing the testing, the tester shall be various, and the testing plan shall be strictly carried out. Moreover, the right testing case shall be chosen.

3) Testing implementation: the testing results shall be recorded in order to be analyzed.

4 After the acceptance testing, the testing team shall analyze the indicators of the testing, and prepare the testing analysis report used as the basis of system review and acceptance. The testing plan and testing analysis report shall be compiled according to the requirements specified in Table F. 0. 5 and Table F. 0. 6 of Annex F.

12. 6 System Operation and Maintenance

12. 6. 1 When the system maintenance is in process, the function structure of system shall be modified and the quality of system shall be improved in accordance with the consumers' comments or variation of external environment, in order to make the system more stable. And the following requirements shall be complied with:

1 The maintenance for error correction shall have a definite stipulation for fault response time and fault removal time. Put forward emergency solutions for faults which cannot be done with in the specified time.

 2 The maintenance for perfection and adaptation shall have a definite stipulation for the principles and process of maintenance as well as conditions and procedures of maintenance implementation.

 3 The maintenance for hardware shall include daily management and maintenance of machinery equipment, management and maintenance programs of corresponding support equipment.

12.6.2 Update, maintenance and backup of data shall be carried out according to the following requirements:

 1 To ensure GIS data is the most recent, the spatial data and attribute data including topographic maps, thematic maps and so on shall be updated in time.

 2 The data should have real-time maintenance. Gradually establish a strict procedure from information collection, entering to review in order to realize the standardization and institutionalization of data maintenance.

 3 In order to ensure reliable recovery of data, not only the basic documents, such as the operating system, database management systems, application software and so on, but also other data files shall be backup.

12.7 Relevant Data Compilation and Submission

12.7.1 After finishing a project, the following information shall be organized:

 1 The demands analysis report.
 2 Overall design.
 3 Database design.
 4 Detailed design.

  5 Testing plan, testing analysis report.

  6 The project summary report whose content and format shall comply with the requirements in Table F. 0. 7 of Annex F.

  7 Software system and database.

12.7.2 After finishing a project, the following information shall be handed in:

  1 The demands analysis report.
  2 Overall design.
  3 Database design.
  4 Detailed design.
  5 Testing plan, testing analysis report.
  6 The project summary report.
  7 Software system and database.

13 Editing and Warehousing of Spatial Data

13.1 General Requirements

13.1.1 Spatial data includes DLG, DEM, DOM, DRG, thematic map data, etc.

13.1.2 The scale of spatial data should be 1 : 500, 1 : 1000, 1 : 2000, 1 : 5000, 1 : 10000, 1 : 25000, etc.

13.1.3 Before editing and warehousing of spatial data, some data shall be collected used as the main base for the technology design of data warehousing according to the project requirement. The collected data involves materials related to the target database, or those that is specially stipulated for spatial database design in Chapter 12 and Section F.3 of Annex F.

13.1.4 The source data shall be produced and completed based on the related specification, and it can pass through the check of data production quality.

13.1.5 Based on the requirements for building a spatial database, the source data shall be checked before entering a database. For those source data which do not meet the requirements, the data editing, processing and transformation shall be accordingly carried out. The check for the warehousing data shall include the following contents:

1 Integrity of data file.
2 Spatial reference system of data.
3 Format of data file.
4 Geometric accuracy of spatial data.

5 Data specifications such as layering and codes of features, geometric characteristics of features, grid size, etc.

  6 Integrity of spatial data, no omission or duplication.

  7 Framing, blocking or partition of data.

  8 Edge matching check of features and attributes, if necessary, edge matching check shall be carried out for data of different periods and different types.

13.1.6 The spatial reference system of warehousing data should be consistent with that of target database. If not, the spatial coordinate transformation shall be executed for the source data.

13.1.7 When data format of source data is inconsistent with that of target database, format transformation shall be carried out to meet the requirement of database building.

13.1.8 When data warehousing, metadata files shall be generated according to the requirements of database building. The metadata file shall be a text file. Each row record shall be arranged as Metadata Item and Metadata Value. Some descriptive information about the spatial data shall be recorded in metadata files, including data source, data specification, production process, production department, spatial reference system, data quality, edge matching, etc. According to the requirement, mapsheet, feature class or dataset may be set as the unit to prepare metadata files of warehousing spatial data.

13.2 Digital Line Graphic

13.2.1 Before warehousing, the spatial data whose format is not GIS shall be edited according to the following requirements:

  1 Data shall be stored in different classes and different layers. Each kind of features shall have its corresponding code. Classification, layering and codes of features shall meet the design requirements of database building.

2 Features of each layer shall be topologically processed. All data after editing shall meet the rules of topological relations.

3 Based on the design requirements of database building, the attribute table of each feature shall be built. And each attribute field of the attribute table shall be assigned a value.

4 When editing data, relationship between features shall be processed, in order that if being displayed at the same time features of different layers shall have a harmonious relationship.

5 The feature's shape shall be complete. Polygon features shall be closed. Lines cannot have dangles. For directed points and lines their directions shall be correct.

6 Quality control shall focus on the check of positional accuracy, attribute accuracy, logical consistency and data integrality.

13.2.2 The warehousing process of DLG shall include technical preparation, data check before warehousing, data process and transformation, data warehousing and data check after warehousing. And the following rules shall be conformed to.

1 When technical preparation, database, data parameters and type of imported data shall be made sure.

2 Data check before warehousing shall be carried out based on requirements specified in Article 13.1.5. If necessary, the following check shall be completed.

1) Check the integrity of spatial feature expression.
2) Check the type consistency: classification and layering of features, if line features is continuous, if feature codes and geometric accuracy are uniform, etc.
3) Check the topological consistency: topological rela-

tionship, the closed relationship of polygons, etc.

3   Transform the source data which do not meet the requirements of database design.

4   DLG data shall be warehoused in batch and in a unit of mapsheet or block.

5   After data warehousing, test the database and carry out the quality check of warehoused data.

13.3   Digital Elevation Model

13.3.1   When technical preparation, some database design requirements shall be determined, including coordinate system, origin, map scale, projection, grid size, coordinate type, scope and so on. Also, digital elevation model (DEM) data and associated metadata shall be prepared.

13.3.2   Before data warehousing, some checks shall be carried out. The check contents include spatial reference system, data format, integrity, accuracy, resolution, segmentation and edge matching, etc. Accordingly, the check result shall be recorded. For those data which is unqualified rework it while for those qualified warehouse it.

13.3.3   Before warehousing of DEM data and associated metadata, those data that do not meet the requirement of data warehousing shall be processed and transformed.

13.3.4   DEM data shall be warehoused in batch and in a unit of mapsheet or block.

13.3.5   After data warehousing, the database shall be tested and the quality of data shall be checked.

13.4   Digital Orthophoto Map

13.4.1   When technical preparation, some design requirements about image database shall be made clear. DOM data and associated metadata shall be prepared.

13.4.2 Before data warehousing, some checks shall be carried out. The check contents include spatial reference system, data format, integrity, accuracy, image tone, mosaic accuracy, cell size, and subset, etc.

13.4.3 Those data which do not meet the requirement of data warehousing shall be processed and transformed.

13.4.4 Image data shall be put in database based on image engineering information or the national standard interchange format. For the national standard images coordinates and resolution of them shall be automatically obtained, while for other images coordinates of the lower left corner and resolution of them shall be input.

13.4.5 After image data warehousing, image pyramids shall be built. We also shall test the database and carry out the quality check of warehoused data.

13.5 Digital Raster Graphic

13.5.1 DRG may be produced by the method of scanning topographic maps or transforming the DLG data to the raster data. When producing DRG, the following rules shall be conformed to.

   1 When DRG is made by scanning topographic maps, the following requirements shall be met:
   1) Technical preparation: collect and analyze the data, as well as develop the program of technical design.
   2) Operating method: scan and geometrically correct the topographic map, deal with images and make related documents.

3) Quality control: eliminate the possible mistakes or omissions by editing.

2 When DRG is made by transforming the vector DLG to the raster data, the following requirement shall be met:

1) Technical preparation: collect and analyze the information, as well as develop the program of technical design.

2) Operating method: symbolize the DLG data, edit the data, transform the vector DLG data to the raster data and make the related documents.

3) Quality Control: eliminate the possible mistakes or omissions by editing.

3 Other requirements for producing DRG may comply with the stipulation specified in CH/T 1015.4 and the design requirements of database building.

13.5.2 The warehousing process of DRG include technical preparation, data check before warehousing, data process and transformation, data warehousing and data check after warehousing. And the following rules shall be conformed to.

1 When technical preparation, database and data parameters shall be made sure. Meanwhile, the technical design shall be carried out.

2 Before data warehousing, some checks shall be carried out. The check contents include spatial reference system, data format, integrity, accuracy, cell size, framing, data quality, etc. Data which is unqualified shall be reworked while data qualified may be warehoused.

3 Data which do not meet the database design requirements shall be processed and transformed.

4 DRG data shall be warehoused in batch and in a unit of mapsheet or block.

5 After data warehousing, security setting and operational testing shall be implemented and the quality check of waterhoused data shall be carried out.

13.6 Thematic Map Data

13.6.1 The thematic map, produced based on geographical basemaps, focuses on expressing some natural and cultural phenomena, such as engineering environment, water and soil erosion, land utilization, project resettlement information, river channel change, etc. The map may be made in vector format and raster format.

13.6.2 Geometric characteristics, classification and codes of the thematic elements in vector thematic map data shall conform to the rules in professional standard specifications and thematic database.

13.6.3 The accuracy of thematic map data for warehousing shall meet the accuracy requirement of the same scale topographic map and that of the professional standard specifications for thematic maps.

13.6.4 Warehousing of the vector thematic map data may be carried out according to the stipulation specified in Section 13.2.

13.6.5 Warehousing of the raster thematic map data may be carried out according to the stipulation specified in Section 13.5.

13.7 Relevant Data Compilation and Submission

13.7.1 After completing a project, the following information shall be compiled:

　1 The technical design document and the design informa-

tion of spatial database.
  2 Source data.
  3 Warehousing data.
  4 Metadata file.
  5 The technical summary report.

13.7.2 After completing a project, the following information shall be handed in:
  1 The technical design document.
  2 Warehoused data.
  3 Metadata file.
  4 The technical summary report.

14 Acceptance of Achievements and Quality Inspection and Evaluation

14.1 Acceptance of Achievements

14.1.1 This chapter applies to the following types of surveying and mapping projects acceptance:

1 The large scale (relatively large-scale) special commissioned surveying and mapping project.

2 The water conservancy preliminary planning surveying and mapping project in a drainage basin or a region.

3 The surveying and mapping projects which have special acceptance requirements in the comprehensive investigation project contracts of large and medium water conservancy and hydropower projects.

14.1.2 Project acceptance shall be mainly based on the documents such as project contracts or assignment books, technical design documents, the relevant laws, the regulations and so on.

14.1.3 Before the project acceptance, the following conditions shall be met:

1 The contents of the project contracts or the assignment books shall have been completed in accordance with the contracts or task agreements.

2 The project quality is all eligible in project inspection, and the relative mistakes and omissions have been processed on requirements or have approved professional opinions.

3 The project data has been arranged according to the requirements.

4 Other conditions specified in the contracts or assignment

books.

14.1.4 The project acceptance shall be organized by the project principals. The corresponding acceptance group shall be established according to the scale of the project. The acceptance group should be constituted by the project principals, supervision units and other units and experts, and no less than 3 technical experts shall in the team. The experts shall have advanced or above technical titles or appropriate professional qualifications.

14.1.5 Project acceptance should be carried out according to the following procedures:

1 The project undertaking units shall send application reports of acceptance to the principals.

2 The principals approve the application report and determine the acceptance time and place, agenda of acceptance meeting, and so on.

3 Convening acceptance meeting; the field sampling inspection and technical pre-acceptance in advance may be organized when needed.

4 Providing the acceptance reports.

14.1.6 The contents of project acceptance shall meet the following requirements:

1 Check whether the achievements meet the requirements of the documents such as the contracts, the assignment books and the approved technical designs.

2 Check the quality of technical plan, personnel and equipment, implementation of interior and field work, process quality control, documents compilation and analysis of results during the formation of project achievements. In addition, the treatment methods shall be put forward. As for the specific professional contents of acceptance, please refer to Appendix G.3.

3 Evaluate the quality of the products according to the requirements refer to the quality evaluation in this standard; if there are supervision units, the project quality evaluation shall be carried on by supervision units. Acceptance meeting shall approve the evaluation results of supervision units.

4 Put forward the handling opinions to the problems found during the acceptance.

5 Prepare the acceptance reports.

14.1.7 Required documents of project acceptance should include both of achievement documents and backup documents:

    1 The results shall include the following:

        1) Documents such as contracts, assignment books, technical designs, etc.

        2) Technical summary report and inspection report.

        3) The control achievement tables of the project, various kinds of drawings, video data, achievement analysis reports, etc.; supervision work reports shall be provided if the project have supervisors.

        4) Equipment installation, as-constructed drawing, description of station, etc.

    2 The backup documents shall include the following contents:

        1) The original records such as observation recorders, satellite images, aerial films or photographs, etc.

        2) The intermediates of achievements in the process of generating calculation results, charts and other products; all kinds of thematic achievements and thematic maps.

        3) The process quality control documents; the calibration data of equipments and instruments.

4) The supervision documents shall be provided if the project has supervisors.

5) Other documents such as communication documents, design change files and other files.

14.1.8 Acceptance report shall be agreed by greater than two-thirds members in the acceptance group. Acceptance group members shall sign the document of acceptance reports. The opinions of objections or reservations held by members shall be specific recorded in the acceptance report documents. The problems shall be amended and improved according to the opinions, and left work shall be completed and data storage shall be implemented. If the achievements have significant mistakes and omissions, contractor shall deal with the problems to meet the requirements according to the opinions.

14.2 Quality Evaluation

14.2.1 The quality grades of project achievements shall be excellent, good, qualified, unqualified four level evaluations. As for the grading standards, refer to Table 14.2.1.

Table 14.2.1 Quality grade standard

NO.	Grade	Score range	Quality to unit ratio (%)			
			Excellent	Good	Qualified	Unqualified
1	Excellent	score$\geqslant$90	$\geqslant$50	$\geqslant$95	100	0
2	Good	90$>$score$\geqslant$75	$\geqslant$10	$\geqslant$90	100	0
3	Qualified	75$>$score$\geqslant$60	—	—	100	0
4	Unqualified	score$<$60	—	—	$<$100	$>$0

14.2.2 When the individual item proceeds to implement the surveying and mapping project quality evaluation in accordance with scoring method, it shall be scored according to the standards of Table G.2.1 to Table G.2.5 in Annex G. The items which are not involved in the table may be executed according to the similar score table.

14.2.3 If the supervisors during the processes are specified, or the surveying and mapping project achievements are divided into several units and the project achievements qualities are evaluated according to the evaluated results of units, the surveying and mapping project quality shall be evaluated by good rate of units. The unit division should be divided combining with the project characteristics and operational requirements as below:

1 The GNSS points, triangulation points, traverse points and the leveling sections at different levels in control survey achievements shall take the "point" or "survey segment" as the unit.

2 For photograph control survey, aerotriangulation and annotation achievements, take the "block of flight strips" or "scene" as the unit.

3 The achievements of topographic maps or image plan maps of different scales, such as surveying and mapping digital topographic and land type maps, cross-section diagrams, etc., shall take the "sheet" as the unit, and the cross-section shall take "strip" as the unit.

4 Deformation monitoring shall take "point" or "period" as the unit.

5 Other project quality units may be classified according to the above methods.

14.2.4 Quality evaluation for comprehensive surveying and mapping projects shall be implemented according to the following requirements:

1 The comprehensive surveying and mapping project is divided into three sub-items of the project technical design, professional surveying and mapping and achievement data compilation and submission.

2 The quality evaluation of the comprehensive survey and mapping projects shall be implemented after the evaluation of the quality of the sub items.

3 The comprehensive quality of the projects shall be evaluated to disqualification when the qualities of sub-items are unqualified. The quality of the sub-items shall be evaluated to disqualification when the qualities of units are unqualified.

4 The score of the project achievement quality, the quality of the sub-items and the quality of the units shall be implemented according to the following requirements:

1) Comprehensive project shall be calculated and scored according to the Table 14.2.4.

Table 14.2.4 Comprehensive surveying and mapping project achievement quality score table

Project sub-items		Weight (P)	Score (Centesimal system)	Remarks
Project technical design		P_1		
Professional survey and mapping sub-items	Horizontal control survey	P_2		
	Vertical control survey			
	Digital topographic survey			
	Cross-section survey			
	⋮			
Results data compilation and submission		P_3		
Total score of project quality (S)		1.00		

2) The surveying and mapping sub-items achievements shall be implemented according to the scoring requirements of individual event in Article 4.2.2.
3) If the sub-items achievements or quality evaluation units of comprehensive projects are all qualified, the quality score S of the project achievements or the sub-items achievements shall be calculated by the weighted average method with Equation (14.2.4):

$$S = \sum_{i=1}^{n}(S_i P_i) / \sum_{i=1}^{n} P_i \qquad (14.2.4)$$

Where:

$S =$ the score of the project achievements or the sub-items achievements

$S_i =$ the score of the project sub items or quality evaluation units.

$P_i =$ The weight of the corresponding sub items or quality evaluation units.

$n =$ The number of the project sub items or quality evaluation units.

4) The weight of each item in the surveying and mapping project achievement quality score table and each surveying and mapping sub-items in the professional items, shall be determined by their importance in the project and proportion of the project amount. The total reference value of P_2, the weight of professional surveying and mapping sub-items should be 0.7. The specific parameter may be adjusted according to the number and importance of units in the professional items in the surveying and mapping projects.

14.3 Quality Inspection

14.3.1 The quality inspection of surveying and mapping achievements includes the process inspection and final inspection. The process inspection shall be completed by the professional survey team or the project department, and the final inspection shall be implemented by the quality management department of the operation unit.

14.3.2 Quality inspection methods of surveying and mapping achievements are divided into full inspection and sampling inspection according to the number of samples examined, and divided into detailed inspection and generalized inspection according to the contents and characteristics of surveying and mapping achievements.

1 Sampling inspection is the method of inspecting the sample unit achievements one by one, if necessary, the important inspecting items which are on the outside of the sample unit results may be inspected generally. Sampling inspection shall extract all relevant documents of sample achievements from the batch of results according to the number of samples. As for the project documents such as project designs, professional designs, supplementary provisions of production processes, technical reports, inspection reports, inspection records, instrument calibration certificates, the copies of the inspection documents and other required documentation, 100% of the original samples or the copies of the samples shall be extracted.

2 According to the requirements of relevant regulations, technical standards and technical designs, the results detailed inspection shall be the method of inspecting the achievements of quality evaluation units one by one and counting the existing various mistakes on the basis of the quality evaluation units and the check items of every sub-item of surveying and mapping achievements, and the quality the sub-items of surveying and mapping achievements shall be evaluated according to the relevant requirements of Section 14. 2.

3 Achievement generalized inspection is the method mainly inspecting quality requirements or indicators which affect the quality of the achievements, systematic bias and mistakes, or tendentious problems. It should record the mistakes and omissions of vital inspection items, critical inspection items and general problems. If the mistakes and omissions of vital inspection items are not found in generalized inspection, the achievement generalized inspection shall be qualified while it would be unqualified if the mistakes and omissions are found.

14.3.3 Process inspection of surveying and mapping achievements shall be the full inspection. The final inspection should adopt the full inspection. The results refer to field inspection items may adopt sampling inspection, and the sampling ratio should be no less than 5% of the total samples while the results outside of the samples shall adopt full indoor inspection.

14.3.4 Before the process inspection of the surveying and mapping achievements, the following conditions shall be met:

　1 All of the work which shall be submitted to the process inspection has been completed.

　2 The mistakes and omissions found in the processes of self-checking and inter-checking by job group have been processed according to the requirement.

　3 Surveying and mapping process documents have been arranged according to the requirements.

　4 Other specified conditions.

14.3.5 Before the final inspection of the surveying and mapping achievements, following conditions shall be met:

  1 All of the work required by the project has been completed.

  2 The mistakes and omissions found in the process inspection have been processed according to the requirement.

  3 The writing of technical summary reports and the arrangement of survey data have been completed according to the requirements of the projects.

  4 Other conditions specified in the contracts.

14.3.6 The process inspection of surveying and mapping achievements shall be implemented on the basis of the surveying and mapping sub-item results for the self-checking and inter-checking of job groups. The mistakes and omissions found in the checking and review results shall be recorded in the surveying and mapping achievements process inspection tables timely, completely, standardly and clearly. After signature of the inspection personnel, treatment personnel and review personal, the process inspection tables of surveying and mapping achievements are forbidden to change, add or delete records.

14.3.7 The final inspection shall be implemented on the basis of the unit results which have passed the process inspection, and process inspection record shall be audited, and the found problems shall be handled as achievement mistakes and omissions. The mistakes and omissions and results of reviews shall be recorded in the surveying and mapping achievements final inspection tables or proofreading and review cards timely and completely. The unqualified results of final inspection shall be returned to the operation department, and the final inspection shall be made again after the operation department finishes its processing. The process is repeated until the inspection is passed. The results shall be submitted to acceptance if all the mistakes and omissions found in the final inspection have been corrected and passed the review. The inspection reports of surveying and mapping achievements shall be written after the final inspection.

14.3.8 In the surveying and mapping results inspection, as for the process inspection tables and inspection report formats of surveying and mapping achievements, refer to Section G.3.

Annex A Horizontal Control Survey

A.1 Drawn the Station Description of Horizontal Control Point

A.1.1 For each horizontal control point, a page of description of station is drawn, as illustrated in Figure A.1.1.

Description of station of horizontal control point

Name of surveyed area

Name and number of station	\multicolumn{4}{c}{Dashan}	Location	Village County Province			
Rough map of position	\multicolumn{4}{c}{[sketch map: To Shancheng, North, Dashan, To Zhangcun, To Yongan]}	The directions of this point to relevant points	Name of point	Order	Distance (km)	Magnetic azimuth (°)
						Jianshan	Order 5	4.5	35
						Zhangjiawan	Order 5	5.0	47
						Wuling	Order 4	7.0	120
						Yangshan	Order 4	7.5	180
						Ligang	Order 3	7.3	263
Type of monument	Order 5 concrete monument	Nearest water source	\multicolumn{6}{l}{Wangzhuang village, 140m southwest of this point}
		Sand source	\multicolumn{6}{l}{Ditch, 200m west of this point}
		Gravel source	\multicolumn{6}{l}{Around Mine hole, 20m southeast of this point}
Section map of buried monument	\multicolumn{5}{c}{[diagram of buried monument section, Unit: cm]}	Siting	Name	\multicolumn{2}{l}{LI××}
							Time		
						Burying monument	Name	\multicolumn{2}{l}{ZHANG××}
							Time		
Note	\multicolumn{9}{l}{Station whether it is selected on an old monument, and which affiliation setups this old monument, and whether it is needed to rebury new monument.}

Figure A.1.1 The format of station description of horizontal control point

A. 1. 2 The name of project should be taken as the name of surveyed area, and the section drawing of monument shall be drawn according to the actual size of buried monument.

A. 2 The Station Description of GNSS Control Point

A. 2. 1 For each GNSS control point, a page of station description is drawn, as illustrated in Figure A. 2. 1.

The station description of GNSS control point

Name of network: Date of filling the table: Date

Name of point:	Numeber of point:	Order:	Number of map:
Rough position: $B=$	$L=$		$H=$
Location:		Authorized depository:	
Nearest residence:		Communication facilities:	
Nearest water source:		Power supply situation:	
Categories of earth:		Gravel source:	
Road map		Traffic conditions	
Rough map of position		Statement of position	
Section map of buried monument		Arranged position of receiver antenna	
Affiliation of burring monument: People who burring monument:		People who selects position:	

Figure A. 2. 1 The format of description of station of horizontal control

A. 2. 2 The name of project should be taken as the name of network, and the section map of monument shall be drawn according to the actual size of buried monument.

A. 3 Specification of Burring Monument of Control Point

A. 3. 1 Buring monument of horizontal control points of the second, third and fourth class shall be implemented according to the following requirements:

1 Monument of horizontal control points of the second, third and fourth class is made of metallic material, and its specification is illustrated in Figure A. 3. 1 - 1.

2 Horizontal control point of the second, third and fourth class may adopt common concrete monument and bedrock monument, and its specification is illustrated in Figure A. 3. 1 - 2 and Figure A. 3. 1 - 3.

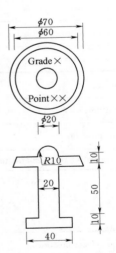

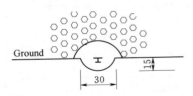

Figure A. 3. 1 - 2 Concrete common monument (Unit: cm)

Figure A. 3. 1 - 1 Metal monument (Unit: mm)

Figure A. 3. 1 - 3 Rock monument (Unit: cm)

A. 3. 2 Specification of buring monument of fifth class control point is illustrated in Figure A. 3. 2.

A. 3. 3 Type of monument and requirements of topographic con-

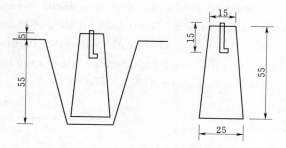

Figure A. 3. 2   Class 5 control point (Unit: cm)

trol point shall be implemented according to the following requirements:

1 For ordinary topographic control point, wooden stake with the size of 5cm × 5cm × 50cm is driven into soil, the stake top is exposed 5cm above the ground, and iron nail is driven into the center of stake top as a sign of mark, with point name written using red paint. If the point position is laid on firm rock, a cross line of 5cm is carved with point name written using red paint.

2 Fixed topographic control points may be buried with monument, or natural rock point, and shall in accordance with the following requirements:

 1) The monument shall be buried with 40cm deep in to the earth, and the exposed section shall be 10cm above the ground, and its surrounding shall be compacted. Type of monument is illustrated in Figure A. 3. 3 - 1.

 2) Natural rock monument may be divided into two kinds: for the first one, a squared holewith the dimension of 15cm × 15cm × 20cm is engraved on the rock, and metallic monument center is poured with concrete, and the name of point is printed on the surface of monument; for the second one, a cross line

(5cm long, 1cm wide, and 1cm deep) is engraved on the center of point position, and 15cm×15cm squared frame line is engraved on the surface of monument and the line groove is 1-2cm deep. Type of monument is illustrated in Figure A. 3. 3 - 2.

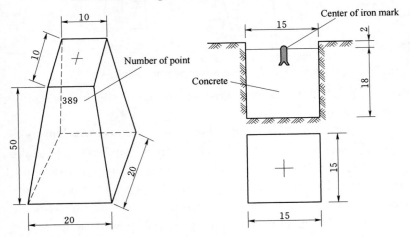

Figure A. 3. 3 - 1 Concrete monument (Unit: cm)

Figure A. 3. 3 - 2 Natural rock monument (Unit: cm)

Annex B Vertical Control Survey

B.1 Bench Mark Diagram

B.1.1 Bench marks may be divided into metal marks and rock marks.

B.1.2 Style and specification of metal mark are as shown in the Figure B.1.2 - 1. Style and specification of rock mark are as shown in the Figure B.1.2 - 2.

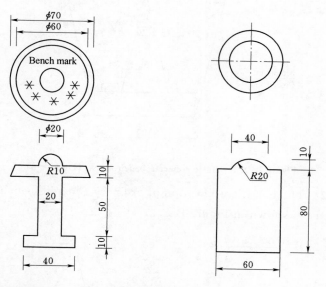

Figure B.1.2 - 1 Metal mark
(Unit: mm)

Figure B.1.2 - 2 Rock mark
(Unit: mm)

Note 1: The spherical part of the metal mark shall be made of copper or stainless steel. The disk and root collaterals may be made of common steel materials.

Note 2: Figure B.1.2-1, Figure B.1.2-2 are bench marks placed on concrete monuments. Figure B.1.2-1 is the bench mark placed on concrete monument, rock monument. Disc diameter of steel pipe stone bench mark shall be determined by diameter and thickness of the pipe. The disk shall be firmly inlaid.

Note 3: "×××××" in Figure B.1.2-1 is the name of a surveying and mapping unit.

B. 2 Bench Mark Stone Specification

B. 2. 1 The burial specification of basic concrete bench mark stone is as shown in Figure B. 2. 1.

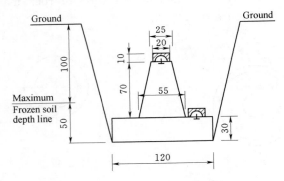

Figure B. 2. 1 Basic concrete bench mark (Unit: cm)

B. 2. 2 The burial specification of ordinary concrete bench mark stone is as shown in Figure B. 2. 2.

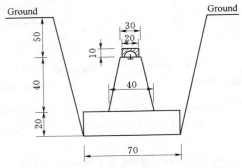

Figure B. 2. 2 Ordinary concrete bench mark (Unit: cm)

B. 2. 3 The burial specification of ordinary rock bench mark stone is as shown in Figure B. 2. 3. When the basic rock bench mark stone is buried, two marks including the upper mark and the lower mark shall be buried. The upper mark and the lower mark may be buried on hard rock with the distance of 0. 5m and the height difference of 0. 1m.

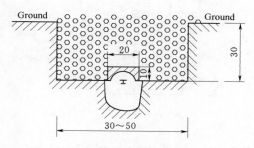

Figure B. 2. 3 Bedrock ordinary bench mark stone burying diagram (Unit: cm)

B. 2. 4 The burial specification of bench mark in the corner is as shown in Figure B. 2. 4.

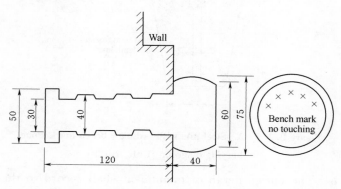

Figure B. 2. 4 Bench mark stone in the corner burying diagram (Unit: mm)

B. 2. 5 The reinforced concrete pile with ordinary bench mark stone shall be buried in the permafrost area as shown in Figure B. 2. 5 - 1. It may be placed by steel pipe bench mark stone in difficult access permafrost area as shown in Figure B. 2. 5 - 2.

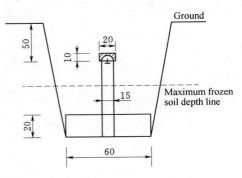

igure B. 2. 5 - 1　Reinforced concrete pile with ordinary bench mark stone burying diagram (Unit: cm)

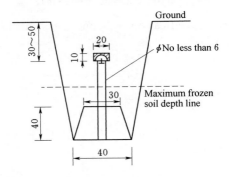

Figure B. 2. 5 - 2　Steel pile with ordinary bench mark stone burying diagram (Unit: cm)

B. 2. 6 The buried depth of monument shall accord to the underground water level and meet the requirements specified in Table B. 2. 6.

Table B. 2. 6   Monument buried depth in the permafrost area

Unit: m

Depth from underground water level to the ground	Depth from the bottom of the monument chassis to the maximum frozen soil depth line	Height from mark to the ground
<6	>0.5	0.3 – 0.5
6 – 10	>0.2	0.3 – 0.5
>0	The burial of concrete ordinary bench mark stone accord to the requirements in general area.	

B. 2. 7   The requirement of top surface of bench mark stone is as shown as Figure B. 2. 7.

Figure B. 2. 7   The top surface of bench mark stone

Annex C　Aerial Photogrammetry

C.1　The Requirements for Layout of Ground Mark Points

C. 1. 1　The photo area, where the ground marks are arranged shall be declared when the contracts of aerial photogrammetry are signed.

C. 1. 2　The ground marks shall be laid out before the aircraft enters the work area.

C. 1. 3　The ground marks should adopt the forms and sizes shown in Figure C. 1. 3. In the figure, $a=0.04M_{pi}$, and the unit

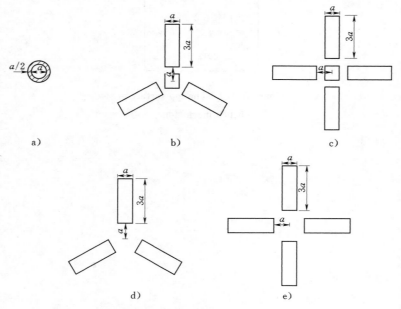

Figure C. 1. 3　The sketch of ground marks

is mm, and M_{pi} is the denominator of the photograph scale. The dimension of the mark center shall not be greater than a, and the width of the wing of the mark and the distance from the wing to the mark center should be a, and the length of the wing of the mark should be $3a$.

C. 1. 4 The colors and material of the ground marks shall meet the requirements as below:

    1 The color contrast between the signs and the ground nearby shall be maximized. White marks shall be laid out on the dark background scene. White marks shall be laid out on the green vegetation while the yellow marks may be adopted too. White signs with black edges should be adopted on the concrete roofs, threshing floors, dirt roads and the grounds without vegetation. For the colors of circular marks, the contrast of colors between the inner ring and the outer ring shall be maximized, at the same time, the contrast of colors between the signs and the ground shall be maximized.

    2 To select the mark material, the following factors shall be taken into consideration: the hue, the conveniences of carriage and installation, easy fabrication, the safety and the price of the material, etc. For example, the paint may be used on the concrete pavement and asphalt pavement while the milky plastic cloth, reed or bamboo matting coated with paint, lime and black cinder may be used on the normal ground.

C. 1. 5 The distribution of marks shall be implemented in accordance with the requirements as below:

    1 The point locations of marks shall be located on the clear targets which are easy to find, such as road intersections, threshing floors and the ends of a dam or a bridge, and the white marks with black edges may be used on those circumstances. If the locations of the chosen points are not easy to find, the marks with wings should be preferred and the three-wing marks should be adopted. The cross marks shall be used when marks are laid out on the control points with targets.

2 When the marks are laid out, the cross points of centerlines of wings or the center of the circular marks shall be overlapped with the locations of chosen points in field or the centers of the existing control points. The wings shall be horizontal in general.

3 When the marks are laid out in city, mining and construction area and shade area, the view angles to air of marks shall be taken into consideration so that the marks in the dead angle of photography can be avoided.

C. 2 The Finishing Format of Papery Controlling Photographs

C. 2. 1 In the papery photograph for pricking points, the red fractions shall be used for annotation of the name, numbering and elevation of points, where the names or numbers of points shall be the numerators, and the elevations of points shall be the denominators. The obverse finishing formats of papery photographs for pricking points are shown in Figure C. 2. 1 - 1. The relevant symbols shall be used of the reverse of photograph to mark the locations of points, where the names or numbers of point shall be marked, and sketchy maps of detailed points locations shall be shown in partial enlarged view to briefly describe the locations of pricking points, the relative altitudes, the engineers for pricking points, the inspectors, the signatures and dates. The textual description shall be concise and precise, and the sketchy locations of points, the textual description, and the pricking points shall be matched to others. The finishing formats of the reverse of papery photograph are shown in Figure C. 2. 1 - 2.

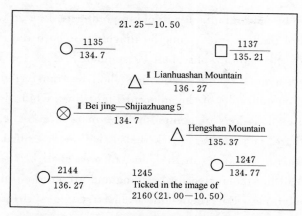

Figure C. 2. 1 - 1 The obverse finishing format of the papery photograph for pricking points

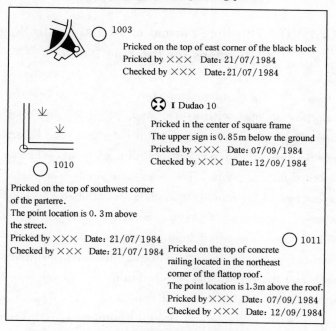

Figure C. 2. 1 - 2 The back side finishing format of the papery photograph for pricking points

C. 2. 2 The photographs for pricking points shall be groomed in the control photographs, and the finishing of both obverse and reverse of the photographs shall be standard, clear and limpid. The public points among flight lines shall be transferred to the principle photographs of the adjacent flight lines while the numbers and explanations of points shall be marked and the photograph where the points are pricked also shall be described. When the photo control points in the adjacent area shall be used, the points shall be transferred for pricking and groomed according to the stipulations. For the points transferred for pricking, the numbers of adjacent photographs and original photographs for pricking points shall be noted.

C. 3 The Finishing Format of Digital Controlling Photographs for Pricking Points

C. 3. 1 In the digital photographs of pricking points, the complete graph, images with 100% and 300% of amplified ratio shall be produced. The maps of point locations shall be arranged in the relevant cells for sketchy location of point, sketchy map and detailed map of points. The relevant contents shall be filled in the cells of name or number of points, coordinates, elevations, operators, inspectors, signatures and dates. In the cell of remarks, the locations of pricking points and relative altitudes shall be described in brief. The textual description shall be concise and precise, and the three types of location maps and the textual description shall be matched to each other.

C. 3. 2 The finishing format of description of station for photo control points by using the film aerial camera images is shown in Figure C. 3. 2.

Description of station for photo control point

No.	P1293	Scale	1/500000	Date of shooting	Mar, 2003
Ticked by	×××	Checked by	×××	Date of pricking	Sep, 01, 2006
Coordinate	X (m)		Y (m)		H (m)
	3375642.05		531356.108		23.04114
Sketchy location of point (No. 0491)					
Sketchy map (100%)			Detailed map (300%)		
Remarks	The point No. P1293 has been pricked at the southeast of pump house, and the elevation has been measured at the top of the house corner.				

Prepared by: Checked by:

Figure C. 3. 2   The finishing format of description of station for photo control points by using the film aerial camera images

C. 3. 3 The finishing format of control point description of station on the digital aerial camera images, such as DMC, UCD, etc. is shown in Figure C. 3. 3.

Description of station for photograph control points

No.	P1293	Scale	1/8000	Date of shooting	May, 2007
Pricked by	×××	Checked by	×××	Date of pricking	Sep, 01, 2006
Coordinate	X (m)		Y (m)		Z (m)
	3375122.05		531226.108		23.04114

Sketchy location (No. 101293)

Sketchy map (100%)

Detailed map (300%)

Remarks: The point No. P1293 has been pricked at the intersection of the side of cart road and the side of embankment slope, and it elevation is as same as that of road.

Prepared by: Checked by:

Figure C.3.3 The finishing format of the description of the control points on the digital aerial camera images, such as DMC, UCD, etc.

C. 4 The Finishing Format of Photographs for Annotation

C. 4. 1 The finishing of photographs for annotation shall be implemented in accordance with the requirements as below:

    1 Sheet number shall be noted on the middle top of the annotation photograph, and the photo number shall be noted on the upper right corner of the annotation photograph.

    2 Borderline for annotation area shall be in blue, while borderline for free margin and the boundary shall be in red.

    3 The right and the bottom lines of the edge matching lines shall be straight, and the left and the top lines shall be the

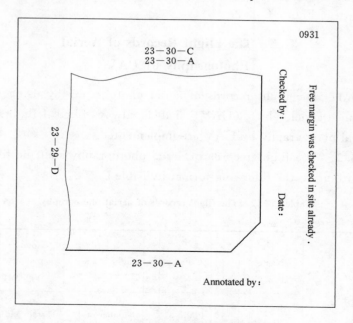

Figure C. 4. 2 The finishing format of photograph for annotation

curve. The number of edge matching shall be noted outboard of the lines.

4 The finishing of annotation shall be implemented in the specified stipulations of diagrammatical symbols. The fair drawing shall be done with different colors. Elements for surface features and lettering shall be in black. Geomorphology elements and remarks shall be in brown. Water elements and remarks shall be in green. The remarks for land boundary and the width of eaves shall be in red.

5 The signatures are necessary for the operators and inspectors.

C. 4. 2 The finishing format of photograph for annotation is shown in Figure C. 4. 2.

C. 5 The Flight Records of Aerial Photography by UAV

C. 5. 1 The flight records of aerial photography by using unmanned aerial vehicle (UAV) shall be always filled if flights of aerial photography by UAV are implemented.

C. 5. 2 The flight records of aerial photography shall be filled according to the following format in Table C. 5. 2:

Table C. 5. 2 The flight records of aerial photography

Aircraft crew _____ Date: _____ Duration ____

Block	Name	Code	Flight block	Ground simpling distance	
	Absolute flying height	Flight direction	Number of routes	Terrain and topography	

Table C.5.2 (Continued)

UAV	Model		ID		Navigation instrument		
Aerial photographic camera	Model		ID		Lens number	Focal length	
	Filter		Aperture		Time of exposure	Sensitivity	
Image	No.				Photography time		
	Pre-test of image				Post-test of image		
Weather	Weather condition		Horizontal visibility		Vertical visibility		
Crew	Operator		Ground station personnel		Photo surveyor	Machinist	
The figure of the flight route							
Remark:							

Prepared by:　　　　Send by:　　　　Received by:

C. 6 The Production Requirements for Control Point Marks of Terrestrial Photogrammetry Photographs

C. 6. 1 Target marks shall be arranged on the chosen photo control points. The targets shall be laid out on a sight rod or a bamboo pole, which shall be upright and firm. The elevations of the mark center shall be measured with the precision of millimeter. The color contrast between the marks and background shall be intense enough.

C. 6. 2 The shapes and types of the target marks are shown as a), b) and c) in Figure C. 6. 2. They may be painted on the cliff or wall in red and white. The four types shown in d), e), f)

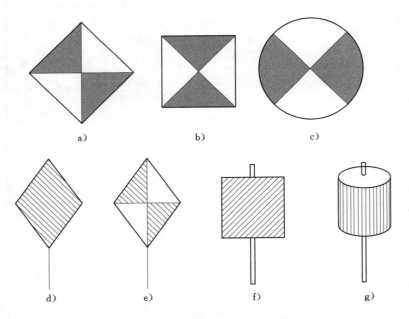

Figure C. 6. 2　The marks on the photo control points

and g) may be painted on wood or asphaltic felt in red and white. The front side of marks shall be directed at the photography station.

C. 6. 3 The size of target marks: the image of marks in the photograph shall be clear and greater than 10 pixels × 10 pixels. Outstanding surface features may be regarded as marks too, but the explanations with drawings shall be declared on the reverse side of photograph.

Annex D   Remote Sensing Image Interpretation

D. 0. 1　The mapborder decoration style of standard framed maps shall be performed as Figure D. 0. 1.

D. 0. 2　The mapborder decoration style of other nonstandard framed interpreted maps shall be performed as Figure D. 0. 2.

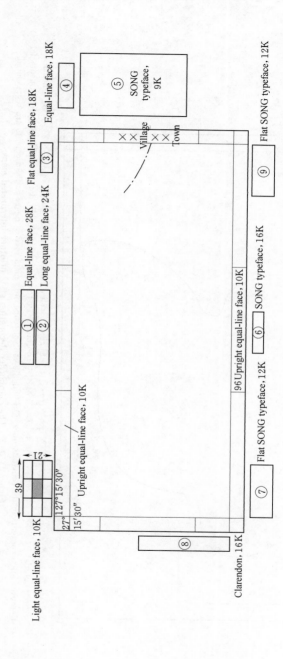

①—Map title; ②—Map number; ③—Security class; ④—Legend; ⑤—Legend content; ⑥—Scale; ⑦—Date for aerial photography or date for satellite imagery acquisition (××××year××month), interpretation date (××××year××month). ×××× coordinate system. ××××height datum. ⑧—Publishing department; ⑨—Inspector ×××, interpretor ×××, cartographer ×××.

(Note: in formal specifications which marked in the small pane on the map is the font and line attributes.)

Figure D.0.1 The mapborder decoration style of standard framed maps

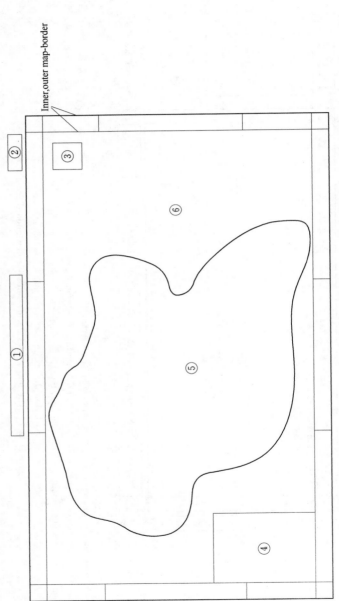

①—Map title; ②—Security class; ③—North Arrow; ④—Legend, scale, imagery acquisition date (×××× year×× month), interpretation date (×××× year×× month), ×××× coordinate system, ×××× height datum, publishing department; ⑤—Representation contents in the control scope; ⑥—Out the control scope only main roads, rivers and settlements are represented.

Figure D. 0. 2　The mapborder decoration style of other interpreted maps

Annex E Special Engineering Survey

E.1 Boundary Monuments of Flooded Line

E.1.1 Figure E.1.1-1 to Figure E.1.1-6 show the types of boundary monuments and non-boundary monuments of line, and the unit is cm.

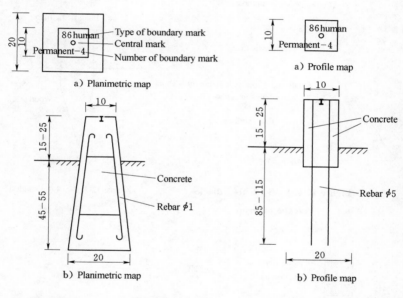

Figure E.1.1-1 Reinforced concrete pile

Figure E.1.1-2 Steel-pipe pile

E.1.2 In order to identify the setting time and types of survey boundary monuments, the number may be consist of setting year, monument type and monument number. For example, the number of residence relocation line, land expropriation line and land utilization line surveyed in 1986 are 86 human-12, 86 land

-12 and 86 use -19. The permanent boundary monument shall be marked as permanent like 86 human-permanent 7. Numbering should be designed while preparing the work plan. All types of the numbers should be in order and avoid repetition. If working in groups, the numbering may start at different number segments.

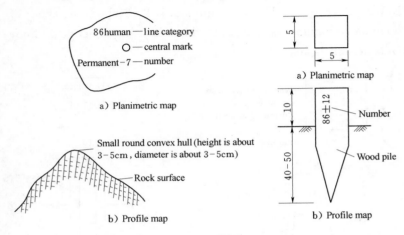

Figure E. 1. 1 - 3 Chisel marks on the outcrop

Figure E. 1. 1 - 4 Wooden stake

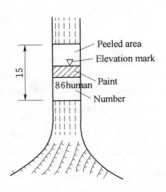

Figure E. 1. 1 - 5 Tree stem mark

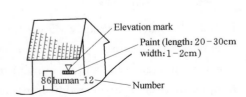

Figure E. 1. 1 - 6 Wall mark

E. 2 Water Area Surveying

E. 2. 1 The GNSS positioning shall meet the following stipulations:

  1 For the real-time differential positioning, the location of reference station shall satisfy the following requirements:

   1) The reference station should be located at the control points with an open vision and the elevation of obstacles in the field of view should be less than 10°.
   2) The reference station should avoid strong magnetic or electrical interference; the distance between the base station and high voltage line, transformer substation or radio transmitting equipment should be not less than 100m, and the distance between the intense radiation transceiver, TV station or microwave relay station should be not less than 500m.
   3) Avoid the object that has strong reflection to the photoelectricity.

  2 The centering error of the reference station antenna shall not over 3mm.

  3 Antennas of moving stations shall be firmly installed at a higher place of the broadside and insulated from metal object. The antenna position should be in the same vertical line with the depth sounder transducer, and the eccentricity correction shall be done if the deviation on the map is larger than 0.3mm.

  4 The elevation of the positioning satellite should be larger than 10°, and the effective satellites for mobile receiver should not be less than 5.

  5 The survey mode, datum parameters, transformation parameters, communication frequency of data link and other parameters of the mobile receiver shall be in accord with the reference station, and the fixed solution shall be used.

6 Before and after the daily sounding survey, the mobile GNSS receiver should be placed on the control point for positioning check. Problems found during the operation shall be examined and compared timely.

7 Positioning data and sounding data shall be synchronous, otherwise the delay correction shall be conducted.

E. 2. 2 The depth sounder inspection shall meet the following stipulations:

1 The depth sounder shall be inspected regularly to check the following items: the stability of the rotation speed, the variation of the zero signal, whether the emission signal and the echo signal are normal, the depth reading difference when change the gain knob and whether all parts of the instrument are working properly. The depth sounder shall be re-checked when it breaks down or the main parts require replacement.

2 The difference between the working and rating voltages shall not exceed 10% for direct current, and 5% for alternative current. The difference between real-time and rating speed shall not exceed 1%, otherwise correction shall be conducted.

3 After the voltage and rotation speed have be rectified, the anchor and navigation on shallow and deep water area shall be checked. If the error appears, the curve of error shall be drawn for correction.

4 Before and after survey, the depth sounders shall be compared on site. The depth sounder may be calibrated by sound velocimeter, hydrophone or inspection plate and the total correction of the depth sounder may be calculated directly based on those calibrations while the depth of water is less than 20m. If the depth is between 20 - 200m, hydrological data may be used to calculate the depth correction, and the error caused by shifting shall be measured.

5 The rope for depth check shall be made by materials with small flexibility and calibrated by steel tape. The depth sounder must be in normal working state when using the inspection board to calibrate it. While the water surface is quiet and the stream is in a low velocity, the inspection depth shall be close to the maximum sounding of the day.

6 When inspecting the depth sounder which may record analog signal and digital signal, both these two signals shall be calibrated, and the analog signal result shall be used as the inspection result.

E. 2. 3 The correction of dynamic draft for sounder transducer may be measured by leveling and GNSS RTK altimetry which shall be in compliance with the following stipulations:

  1 Leveling shall meet the following requirements:

    1) Select an appropriate observation water area with a similar sounding that the survey area has. Set a buoy for the observation point on the water surface and set the level on the shore at a positon with appropriate height.

    2) The leveling staff shall be erected at a proper position in the transducer installation area.

    3) In the relative static state, use the level to read the staff five times and take the average as the static observation value h_1.

    4) Go through the observation point five times continuously at the speed of sounding and use the leveling to read the staff. Take the average as dynamic observa-

tion value h_0. Use the method in Item 3) of this section to calculate a static observation value h_2 again.

5) The correction for dynamic draft shall be calculated according to Equation (E. 2. 3 - 1):

$$\Delta h = h_0 - \frac{(h_1 + h_2)}{2} \qquad (E. 2. 3 - 1)$$

Where:

Δh = correction of dynamic draft, and the unit is m. The correction may be skipped if the correction is smaller than 0.05m.

2 The GNSS RTK altimetry shall meet the following requirements:

1) In the relative static state, use the GPS RTK positioning technology to read the antenna height of moving station five times, and take the average value as static observation value H_1.

2) Go through the observation point five times continuously at the speed of sounding and read the antenna height of moving station. Take the average value as dynamic observation value H_0.

3) The correction for dynamic draft shall be calculated according to Equation (E. 2. 3 - 2):

$$\Delta h = H_1 - H_0 \qquad (E. 2. 3 - 2)$$

E. 2. 4 When multi-beam sounding system is used for scanning survey operations, the operating conditions, equipment installation and calibration, sounding, data processing and mapping shall be performed in accordance with the following stipulations:

1 The operating conditions shall meet the following requirements:

1) When using the GNSS dynamic real-time positioning

system, the positioning accuracy shall be better than ±1m.

2) The operation weather condition shall be better than (include) the second grade sea state (wind force 4, wave height 1m). When the horizontal roll angle measured by the attitude sensor (wave compensator) is more than 8°, or the longitudinal inclination angle is more than 8° the operation shall be stopped. When multi-beam is only used for obstacle detection and the formal sounding map does not need be provided. And the operation may not be restricted by sea state under ensured safety.

3) While the quality of echo signal has been ensured and the noise level is low, the maximum speed of ship shall not exceed the limit value calculated by Equation (E. 2. 4):

$$V = \tan\left(\frac{\theta}{2}\right) H_m n \times 3600 \qquad \text{(E. 2. 4)}$$

Where:

θ = the nadir coverage angle by cross receiving along the travel direction;

H_m = the shallowest depth of a measuring section in the survey area. The average depth may be used in general survey while the water depth changes less, mm;

n = the sample number of multi-beam per second.

4) The main measuring line shall parallel to the general direction of depth contours. The check lines shall be evenly laid in the direction which is vertical to the main measuring line. The check line may use the single-beam or multi-beam depth sounder.

5) The ratio of coverage width of multi-beam and the sounding below the transducer is called the width-depth ratio. The interval between survey lines shall consider various factors such as the observation ship, underwater topography, operation purpose and instrument nominal index etc.

6) Depth sounding shall adopt sound velocity meter to correct the sound velocity. The sound velocity (or sound velocity ϕ vertical – section) shall be measured before and after the sounding. More measurements are needed if there are indications that the sound velocity ϕ vertical – section changes significantly.

2 The installation and calibration of equipment shall meet the following requirements:

1) The system installation layout shall reduce the comprehensive noise to the lowest level, and the bottom installation shall be a priority.

2) The attitude sensors should be installed in the position which may accurately reflect the location of the observation ship or multi-beam transducer, and its direction should parallel to the direction of fore and aft line. The relative position of antenna and attitude sensor shall be determined before operation.

3) The electric gyrocompass shall be installed on the observation ship's fore and aft line and the zero reading shall point to the bow.

4) The average depth of the calibration area shall not be less than the maximum depth of the survey area. If possible, the water area where has been surveyed with reliable data shall be selected.

5) The order of the calibration project is delay, pitch angle, roll angle, and yaw. The calibration should choose independent land features such as the flat slope with a tilt angle greater than 10°, a reef or shipwreck. A flat area shall be chosen for roll angle calibration. The independent land feature such as reef, shipwreck and others shall be chosen for yaw calibration.

6) The key parameters of the multi-beam transmission/reception unit shall be set according to specific conditions of the survey area, observation ship and equipment.

7) The draft of the observation ship in four directions (front, back, left and right) and the draft of transducer shall be measured respectively by two people. The difference value shall not be larger than 0.1m and the average value shall be taken as the final result. The boat squat at different speed from standstill to maximum speed shall be measured.

3 The sounding shall meet the following requirements:

1) In the process of sounding, the running state of the dynamic sensor, electric gyrocompass, positioning and sounding equipment shall be real-time monitored. The operation should be stopped when they break down.

2) The stable time of on-line measurement shall be 1min (on the extension line of survey line and traveling according to the predetermined direction and speed).

3) After the field work, the key parameters of multi-beam system shall be checked again. The original field

data shall be timely converted to the data format which may be used by the software package of indoor data processing.

4 Data processing and mapping shall satisfy the following requirements:

1) The contents of data processing include eliminating the gross error of navigation data and sounding data and data thinning.

2) Check the processed data, and read the three-dimensional coordinate of appropriate sounding points to verify their correct depth. Get the interval distribution of water depth in order to statistically compare the processed water depth.

3) The underwater DTM model shall be built by using system software. The model should accurately reflect the overall trend of underwater topography and every typical topographic point. For large scale survey projects, the DTM model may be built in different partitions.

4) Generate the underwater contours or depth contours automatically, and get the best underwater topographic map by editing and modification.

5) According to the task requirements, cut vertical – section and cross – section from the DTM model to generate section data and underwater section map.

Annex F GIS Development

F. 0. 1 The demand analysis report shall be written as the format in Table F. 0. 1.

Table F. 0. 1 Writing format of requirement analysis report

1 Introduction
1. 1 The compilation purpose: explain the purpose of compiling the requirement analysis report, and point out the intended readers.
1. 2 Background:
1. 2. 1 The name of the system to be developed.
1. 2. 2 The proponent, developer and user of the project.
1. 2. 3 The basic relationship between this system and other systems or other institutions.
1. 3 Definitions: list the definition of special terms appearing in this report.
1. 4 References: list the necessary references.
2 Task overview
2. 1 Target: describe the intention, application goal, action scope and other background information which shall be told to readers about this system development. Also, explain the relationship between the developed system and other related systems.
2. 2 User business description: describe the user's organizational structure, business responsibilities, business terminology, user business flow and the user's specific business needs.
2. 3 User information Status: explain the user's existing information systems, and their health and usage.
2. 4 End-User Features: list end-user features of this system, and fully describe the education level and technical excellence of operators and maintainers, the expected use frequency and the expected end-users number of the system.
2. 5 Assumptions and constraints: list assumptions and constraints of developing the system, such as contract amount, schedule, etc.

Table F. 0. 1 (Continued)

3 Requirement stipulations

3.1 Stipulations for functions: item by item following a list, quantitatively and qualitatively propose the functions the system shall realize.

3.2 Stipulations for data: respectively describe the stipulations for spatial data and non-spatial data.

3.3 Stipulations for performance.

3.3.1 Accuracy: describe the accuracy requirements for input and output data of the system, may include accuracy in the transmission process.

3.3.2 Requirements for time characteristics: describe requirements for time characteristics of the system.

3.3.3 Flexibility: describe the system's flexibility requirement, namely the system's adaptation ability to some changes in user needs.

3.4 Requirements for input and output: explain the type of input and output data, and item by item state its media, format, value range, accuracy and so on. Explain and illustrate the system's data output and control output which shall be labeled.

3.5 Requirements for data management ability: show the number of files and records to be managed, the scale of tables and files. It is necessary to estimate the storage requirements according to the foreseeable data growth.

3.6 Requirements for troubleshooting: list software and hardware failures possibly happening, effects on every performance, as well as requirements for troubleshooting.

3.7 Other special requirements: such as, user unit's requirements for security and secrecy, ease of use, special requirements for maintainability, expandability, readability, reliability and convertibility of operating environment, etc.

4 Stipulations for operating environment

4.1 Devices: list hardware devices needed for the software operation.

4.2 Support softwares: list support softwares, including the operating system, compile program, the test supporting software, etc.

4.3 Interface: describe the interface and data communication protocols between this system and other systems.

F. 0. 2 The overall design report shall be written as the format in Table F. 0. 2.

Table F. 0. 2 Writing format of overall design report

1 Introduction
1.1 The compilation purpose: explain the purpose of compiling the overall design report, and point out the intended readers.
1.2 Project background: describe client, development unit and competent authority as well as the relationshipbetween the software system and other systems.
1.3 Definitions: list the definitions of special terms appearing in this report.
1.4 References.
2 Task overview
2.1 Target: describe the main goal of developing this system.
2.2 Operating environment: list the hardware environment and software environment needed for the software operation.
2.3 Needs summary: describe user's main demands for the system.
2.4 Conditions and restrictions:
2.4.1 Favorable conditions for development.
2.4.2 Unfavorable factors or restrictions for development.
3 Overall design
3.1 The guiding idea.
3.2 The main technical route.
3.3 General structure design: depending on demand, the system is divided into different subsystems, each contains different functional modules; design the general structure of the system.
3.4 Functional design: design the basic functions of every system module.
4 Interface design
4.1 External interface: the user interface design: including screen design, menu design, prompt message design and so on. It requires the surfaces with a unified interface style, simple and clearly targeted. Software and hardware interfaces: design interface modes of the system's hardware and software.
4.2 Internal interface: design interfaces among system modules.
5 Error handling design
5.1 Error message: describe possible errors or troubles during the system operation, including mode, implication and handling method of the system output information.
5.2 Remedy: explain the remedy or solution for system operation troubles.

F. 0. 3 The database design report shall be written as the format in Table F. 0. 3.

Table F. 0. 3 Writting format of database design report

---
1 Introduction

    1. 1 The compilation purpose: explain the purpose of compiling the database design report, and point out the intended readers.

    1. 2 Background: introduce the name of the database to be developed and the name of the software system which will use the database; describe client, development unit and competent authority as well as the relationship between the software system and other systems.

    1. 3 Scope and content involved in the database design.

        1. 3. 1 Scope involved in this design.

        1. 3. 2 Content involved in this design.

    1. 4 Definitions: list the definitions of special terms appearing in this document.

    1. 5 References: as the base of database design, list the data references in the database design, including national standards and professional standards associated with the system database, the system requirement analysis report, etc.

2 Instructions about database environment

    2. 1 Describe the database management system, design tools, programming tools and so on, which used in the database.

    2. 2 Detailed configuration.

3 Content and volume of data

    3. 1 Data content: explain various types of data contained in the database: spatial data, attribute data, system management data, metadata and so on. Determine the format, plane coordinate system, height datum, scale and other things of various types of data.

    3. 2 Data volume: reasonably estimate the data volume of various types of data in the database.

4 Logical and physical design of database

    4. 1 Selection of spatial data model: determine the spatial data model used by spatial data.

    4. 2 Logical Design: describe the types, attributes of the various data entities which are managed by the database. Introduce conceptual models of attribute data and spatial data, as well as relevance of spatial data and attribute data. Determine the logical structure and naming rules of the database. The database to subsets (datasets) shall be subdivided based on different data types. Fundermental geographic information database is logically divided as follows: DLG dataset, DEM

Table F. 0. 3 (Continued)

dataset, DOM dataset, etc. Determine the storage mode of vector data, raster data and attribute data. Fundermental geographic information database shall be designed according to the stipulations specified in CH/T 9005 *Basic Specifications for Fundermental Geographic Information Database*. Thematic database design shall be compatible with the database-related stipulations in the prevailing national standards, provincial standards or professional standards.

 4.3 Physical design: determine the physical structure of the database. Choose the database management system, storage structure and access methodsaccording to the logic structure of database.

5 Layering, classification and coding of data

 5.1 Rules for data layering and naming of layers and files: determine the layer name of the graphic data, rules for data contents and naming rules of files. Naming rules of files comply with the stipulations specified in CH/T 1005 "*Digital Products of Fundamental Geographic Information Naming Rules for Data Files*".

 5.2 Classification and coding of data: determine the category of fundamental geographic information features and thematic information features. Make sure the attribute table contents of features, as well as the order and codes of attribute items. Classification and codes of fundamental geographic information features comply with the stipulations specified in GB/T 13923 *Specifications for Feature Classification and Codes of Fundamental Geographic Information*. Classification and codes of thematic information features comply with the stipulations specified in GB/T 18317 *Classification and Code for Thematic Map Information*. At the same time, it shall be compatible with the stipulations about classification and codes in the prevailing national standards, provincial standards or professional standards. The code extension shall comform to the rule of scientificity, systematicness, expandability and compatibility.

 5.3 Data dictionary: make data dictionary to detailedly describe and define spatial data and attribute data. Data dictionary of fundamental geographic information features comply with the stipulations specified in GB/T 20258 *Data Dictionary for Fundamental Geographic Information Features*. Data dictionary of thematic information features shall be compatible with the stipulations about data dictionary in the prevailing national standards, provincial standards or professional standards.

 5.4 Metadatabase: determine the content of metadata. Metadata design of fundamental geographic information features comply with the stipulations specified in GB/T 19710 *Geographic Information-Metadata*. Metadata design of thematic information features shall be compatible with the stipulations about metadata design

Table F. 0. 3 (Continued)

in the prevailing national standards, provincial standards or professional standards.

5.5 Design of symbol library: contain the design of point, line and area symbols, as well as that of annotation. Design of map symbol library comply with the stipulations specified in CH/T 4015 *General Rules for Building Cartographic Symbols Base*.

6 Summary of tables

Table name	Function description
Table A	
Table B	
⋮	
Table N	

6.1 Table A-Table N.

Table name			
Field name	Data type (accuracy range)	Null/not null	Description
			Field meaning, constraint condition
Additional description			

7 Safety Design

7.1 The method for preventing users directly manipulate the database: The user can only log in with the account to the application software, and access the database only through the application software, no other way to operate the database.

7.2 User account password encryption method: the user account password is encrypted to ensure that not appear anywhere with plaintext passwords.

7.3 Roles and Permissions: determine the operating authorization of the database tables for each role such as create, retrieve, update, and delete. Each role has permission to just be able to complete the task, neither more nor less. Assign roles to users when applying, and each user's permission equal to the sum of all his roles'permissions.

Table F. 0. 3 (Continued)

Role	The tables and lines permit to access	Operating authorization
Role A		
Role B		

8 Instruction of database management and maintenance

  Timely provide methods of the database management and maintenance when design the database, which is helpful to write a correct and integrated user's manual.

F. 0. 4 The detailed design report shall be written as the format in Table F. 0. 4.

Table F. 0. 4 Writting format of database design report

1 Introduction

  1. 1 The compilation purpose: explain the purpose of compiling the detailed design report, and point out the intended readers.

  1. 2 Background: including the name of the system to be developed. List the proponent, developer and user of this project.

  1. 3 The scope and contents of the detailed design:

  1. 3. 1 The scope of this design.

  1. 3. 2 The contents of this design.

  1. 4 Definitions: list the definitions of special terms appearing in this report.

  1. 5 References: list the correlative references.

2 Description of system structure

  2. 1 Naming rules of module: determine the naming rules of the modules to ensure that the style is consistent between module design document and codes.

  2. 2 Summary table of module.

Subsystem A	
Module name	Function brief
⋮	
Subsystem B	
Module name	Function brief
⋮	

Table F. 0. 4 (Continued)

3 Design specification of module 1 (identifier)

3.1 Module description: gives a brief description of the module, mainly explaining the purpose and significance of designing this module. And characteristics, functions and performance of this module shall be also explained.

3.2 Input item: analyze and determine the logical structure of the input data, use charts to depict the structures of these data and give the characteristics of each input item.

3.3 Output item: analyze and determine the logical structure of the output data, use charts to depict the structures of these data and give the characteristics of each output item.

3.4 Algorithm: give a detailed description of the selected algorithm in this program, including the specific calculation formulas and calculation steps.

3.5 Logic flow: use charts (e. g. flow charts, decision tables, etc.) and some necessary instructions to represent the logical flow of this module.

3.6 Interface: explain the logical connection type between this module and other related modules, and the parameter passing mode involved.

3.7 Restriction: describe the restrictions in the operation of this module.

3.8 Unresolved issues: give a description of the issues that has not been solved in the module design but shall be solved before completing the software.

4 Design specification of module 2 (Identifier)

The design mode of module 2 is similar to module 1, and by this analogy to module N.

In the similar way of article third, explain the design thinking of model second and even model N.

F. 0. 5 The test plan report shall be written as the format in Table F. 0. 5.

Table F. 0. 5 Writing format of test plan report

1 Introduction

1.1 Compilation purpose: explain the compilation purpose of the test plan report, and point out the intended readers.

1.2 Background: including the name and intro of the system to be developed. List the proponent, developer and user of this project.

1.3 Definitions: list the definitions of special terms appearing in this report.

Table F. 0. 5 (Continued)

  1.4 References: list the correlative references.

2 Plan

  2.1 Software description: provide a chart as the outline of test plan, term by term declaring the functions, input, output and other quality indicators of the software tested.

  2.2 Test contents: lists the name identifier, test schedule, content and purpose of each test content in integration testing and validation testing, such as module function, interface correctness, running time, design constraint and limit testing.

  2.3 Test 1 (identifier): provide the participants of test contents and the function modules tested.

    2.3.1 Schedule: provide the test schedule, including the test date and the corresponding contents.

    2.3.2 Conditions: state the resources required in this test, including the following ones:

      2.3.2.1 The type, quantity and scheduled time of the equipment to be used.

      2.3.2.2 List the software that will be used to support this test procedure but is not a part of the tested software, such as test drivers, test monitoring procedures and so on.

      2.3.2.3 List the staff number, technology level and related propaedeutics, which provided expectedly by users and development task group during the test.

    2.3.3 Test information: lists the information required in this test, including the following contents:

      2.3.3.1 Documents related to this task.

      2.3.3.2 The tested program.

      2.3.3.3 Input and output cases for testing.

      2.3.3.4 Procedural charts and test method for controlling this test.

    2.3.4 Test training: explain or use materials to explain the training plan for the use of the tested software. Stipulate the contents, trainees and trainer of the training.

  2.4 Test 2 (identifier): the design mode of test 2 is similar to that of test 1, and so the after test.

3 Specification of the test design

  3.1 Test 1 (Identifier): explain the design mode of the first test.

    3.1.1 Control: describe the control mode of the test, for example, the input mode is manual, semi-automatic or automatic, as well as the sequence of controlling operation and method of recording results.

Table F. 0. 5 (Continued)

>   **3. 1. 2** Input: describe the input data and the strategy of selecting these data used in this test.
>   **3. 1. 3** Output: describe the expected output data, such as test results, possible intermediate results or operational information.
>   **3. 1. 4** Process: describe every step and control command for completing this test, including the preparation, initialization, intermediate steps and the method for ending the operation of this test.
>   **3. 2** Test 2 (Identifier): the design mode of test 2 is similar to that of test 1, and so the after test.
> **4** Evaluation criteria
>   **4. 1** Scope: describe the checking scope of the selected test cases and their limitations.
>   **4. 2** Data compilation: explain the transformation and processing methods for making the test data into an appropriate form so that the test results may be compared with the known results, such as a manual method or an automatic one. It is necessary to explain the required hardware and software for the test when using the automatic method.
>   **4. 3** Scale: describe the evaluated scale used to determine whether the test is qualified, such as the type of reasonable outputs, allowable deviation scope between the output result and the expected result, and maximum times of allowable interruptions or machine halts.

F. 0. 6 The test analysis report shall be written as the format in Table F. 0. 6.

Table F. 0. 6 Writing format of test analysis report

> **1** Introduction
>   **1. 1** Compilation purpose: explain the compiling purpose of the test analysis report, and point out the intended reading range.
>   **1. 2** Background: including the name of the system to be tested. List the proponent, developer and user of this project. Point out the differences that may exist between the test environment and the actual operating environment, and how these differences affect the result of the test.
>   **1. 3** Definitions: list the definitions of special terms appearing in this report.
>   **1. 4** References: list the references used.

Table F. 0. 6 (Continued)

2 Summary of the test

List the identifier and contents of each test in the tabular form. Indicate the differences between actual contents and pre-designed contents in the test plan, and explain the reason.

3 Test results and discovery

3.1 Test 1 (identifier): compare the actual dynamic output with the requirements of dynamic output, and state every discovery in the test.

3.2 Test 2 (identifier): the description of the contents and discovery in test 2 is similar to that of test 1, and so the after test.

4 Conclusion to the functions of software

4.1 Function 1 (identifier).

4.1.1 Capability: briefly describe the function, explain the software capabilities designed to meet this function and the ability which has been confirmed by one or more tests.

4.1.2 Restriction: describe the range of the test data value; list the drawbacks and limitations of this function during the test.

4.2 Function 2 (identifier): the test conclusion of functions in Chapter 2 is similar to that of Section 4.1, and so the after chapter.

5 Summary of analysis

5.1 Capability: state the capabilities of this software which have been confirmed by testing. Compare the test results with the required results if the test is used to verify one or more specific performance has been realized. And point out the effects on capability testing caused by the differences between the test environment and the actual operating environment.

5.2 Defect and restriction: state the defects and restrictions conformed by the test, describe the effects on the software performance caused by each defect and restriction, and explain the cumulative effects and total effects of all performance defects tested.

5.3 Suggestion: put forward improved suggestions for each defect, such as the modified methods, degree of urgency, expected workload and responsible person of each modification.

5.4 Evaluation: explain whether the development of the software has reached the intended target or not, and whether could be used or not.

6 Resource consumption for testing

Summarize the data of resource consumption for testing, such as the technology level, rank, and number of the staff, machine-hour consumption and so forth.

F. 0. 7 The project summary report shall be written as the format in Table F. 0. 7.

Table F. 0. 7 Writing format of project summary report

1 Introduction
   1.1 Compilation purpose: explain the compiling purpose of the project summary report, and point out the intended reading range.
   1.2 Background: including the names of the project and the completed system. List the proponent, developer and user of this project.
   1.3 Definitions: list the definitions of special terms appearing in this report.
   1.4 References: list the references used.
2 Actual development results
   2.1 Products: explain the final products developed, including the name of each program in the program system, and the hierarchical relationship among them, the program amount in kilobytes, the form and amount of storage media; how many versions the program system has, and the version number and difference between them; the name of each document; each database established. Compare with the configuration management plan if the plan has been formulated during the development.
   2.2 Main function and performance: list the actual main functions and performance of the software product term by term; according to the contents of the overall design estimate if the development aims have been achieved.
   2.3 Basic flow: using diagrams to depict the actual basic workflow of the system.
   2.4 Schedule: list the contrast between the original schedule and actual schedule, and analyze the main reason of the discrepancy.
   2.5 Cost: list the contrast between the planned cost and actual cost, and analyze the main reason of the discrepancy.
3 Evaluation of the development work
   3.1 Evaluation of production efficiency: evaluate the actual production efficiency and list the contrast between it and the planned efficiency.
   3.2 Evaluation of product quality: explain the occurrence rate of programming error checked out in the test; also mean the number of false commands in one thousand commands.
   3.3 Evaluation on technique: evaluate the technique, methods, tools and means used in the development.
   3.4 Reason analysis of error: analyze the reasons of errors appearing in the development.
4 Experiences and lessons
    List the main successful experiences and lessons which are acquired from the system development work, and suggestions for the future system development work.

Annex G Product Acceptance and Quality Inspection and Evaluation

G. 1 Major Professional Content of Project Acceptance

G. 1. 1 The inspection contents of horizontal control survey, vertical control survey, digital topographic map surveying and mapping, aerial photogrammetry, terrestrial laser scanning and terrestrial photogrammetry, map compilation, special projects such as road survey, transmission line measurement, construction control network survey, measurement of engineering deformation monitoring network, deformation measurement of reservoir bank slope , etc. can be implemented according to the relevant provisions of GB/T 24356.

G. 1. 2 The vertical and cross – section survey and underwater topographical survey of water area survey shall implement the acceptance in accordance with the following contents:

  1 The coordinate system and height system.

  2 The layout, point selection, density, embedding and observation scheme of horizontal and vertical control network.

  3 Underwater topographical survey plan, and accuracy of digital survey achievements.

  4 The working method, arrangement of the sections, scale and accuracy of vertical – section and cross-section survey.

  5 The drawing and plotting error of vertical – section and cross-section, and the mapping scale of vertical – section and cross-section.

6 Water level line, flood level line.

7 The consistency among the number and written notes in the map, the observation handbooks and achievement lists, and the completeness of contents, the correctness of diagrammatical symbols use.

8 The identification of instrument model, class, precision and the mapping software system.

9 Whether the projects were implemented and the technical change according to the requirements of technical design.

10 Whether the data arrangement was completed and normative.

G. 1. 3 For the aero space photogrammetry, the following contents for acceptance inspection shall be checked:

1 The design and survey plan for photo control points.

2 The tolerance and precision of control survey.

3 The rule, method, requirement of annotation and necessary preparation for photo annotation.

4 The block adjustment accuracy, the pass point location and plan. The requirements and technical indicators for data preparation, relative orientation and bridging of model, gross error test and automatic location, adjustment, connection of pass point in adjacent area, model output, etc.

5 The operation methods and technical indicators of orientation modeling.

6 The basic requirements, methods, quality control measures and technical indicators for various kinds of data acquisition.

7 The inspection items and accuracy requirements of various kinds of instruments and equipment.

8 Achievements submission and the requirements of data.

9 Whether the projects were implemented and the technological means were changed according to the requirements of technical design.

10 The conditions of equipment inspection and software identification.

11 The conditions of achievements meeting the requirements of technical design and specification.

12 Whether the data arrangement was completed and normative.

G. 1. 4 The special engineering surveys such as the projects in water basin or region shall implement the acceptance of the following contents:

1 The grading and layering of control network.

2 Precesion and density requirement of control network and implementation conditions.

3 Other contents are as same as conventional horizontal control survey and vertical control survey.

G. 1. 5 The special engineering surveys such as water conveyance line survey and levee engineering survey shall implement the acceptance inspection in accordance with the following contents:

1 The inspection items and contents of horizontal control survey and vertical control survey are as same as the control survey of special engineering surveying in water basin or region.

2 The scope, method, precision and achievement of water conveyance line banded topographic map, dike line banded topographic map, beach topographic map and detention basin topographic map are as same as the digital topographic survey.

3 The method, layout, scale and survey precision of profile survey and cross-section.

4 The drawing and plotting error and drawing scale of vertical-section and cross-section.

5 The Instrument model, class, precision and inspection. The evaluation conditions of mapping software system.

6 Whether the relationship between the newly and previously survey achievements have been explained clearly.

7 Whether the data arrangement was completed and normative.

G. 1. 6 The special engineering surveys such as the land acquisition and resettlement allocation engineering survey shall implement the acceptance inspection in accordance with the following contents:

1 For the topographic and land boundary map survey, the acceptance content shall be corresponding with the relevant operating method and the acceptance inspection of formed achievements shall refer to similar projects.

2 The acceptance inspection of boundary monument survey includes: location, density, type, measurement accuracy of boundary monument layout; The control survey contents of boundary monuments shall be implemented in accordance with the control survey of engineering surveying in the water basin or region. The boundary monument setting off and embedding method; whether each kind of boundary monument is managed, and the detailed description and detailed figure of important boundary points, and whether the plotting and record of permanent and temporary monuments are correct; whether the height system of boundary monument achievements are corresponding with the system of reservoir topographic map and river section survey in the reservoir design phase; whether the data arrangement was completed and normative.

3 The moving measurement contents of control points are as same as the control survey of special engineering surveying in water basin or region.

4 The acceptance inspection of resettlement planning site survey shall refer to the similar products in the above chapters.

G. 1. 7 The geological exploration and survey shall implement the acceptance inspection in accordance with the following contents:

1 The basic work contents.

2 The survey and design method and connective measurement of various kinds of geological prospecting points.

3 The scale, precision and mapping requirements of geological mapping and surveying.

4 Other special measurements, such as connection survey of shaft or adit, geophysical survey, etc. the acceptance inspection contents refer to the above chapters according to the operation methods and the surveying and mapping product types.

5 Whether the data arrangement was completed and normative.

G. 1. 8 GIS development shall implement the acceptance inspection in accordance with the following contents:

1 The basic structure of the system design.

2 The basic functions of the system.

3 The situation of the system response of requirements.

4 The compatibility and expansibility of the system.

5 Testing, operating stability and maintainability of the system.

6 Experts' opinions on the system.

G. 1. 9 The editing and storage of data achievements shall implement the acceptance inspection in accordance with the following contents:

    1 The data quality precision requirements and the basic structure and specification of database, including code classification, data attribute, data format, data dictionary, metadata, etc.

    2 The relative contents of editing, storage and raster graphic storage of digital Line graph (DLG), digital elevation model (DEM), digital orthophoto map (DOM), special data, etc., mainly including data storage inspection, projection transformation, data processing, database building, examination, compiling data dictionary, etc.

G. 1. 10 The project acceptance shall include others contents required by clients or technical design as well.

G. 2 Scoring for Individual Projects

G. 2. 1 As for the evaluation and scoring of surveying and mapping achievement quality of the horizontal and vertical control survey, refer to Table G. 2. 1.

G. 2. 2 The quality evaluation and scoring table of the digital topographic surveying and mapping achievement units shall be implemented according to Table G. 2. 2.

G. 2. 3 The quality evaluation and scoring of water area survey (longitudinal-section survey, cross-sectional survey, underwater topographic survey) shall be implemented according to Table G. 2. 3.

G. 2. 4 The quality evaluation and scoring of boundary markers layout and survey in resettlement engineering survey shall be according to Table G. 2. 4.

G. 2. 5 The quality evaluation and scoring of results data editing and warehousing sees table G. 2. 5.

Table G. 2. 1 The evaluation and scoring of surveying and mapping achievement quality of the horizontal and vertical control survey

Branch quality	Score	Sub-item of quality	Score	Quality unit	Score			
					Excellent	Good	Qualified	Unqualified
Monumentation of mark-stones	30	Point selection	10	Rationality of Point distribution	3.00 – 2.70	2.60 – 2.30	2.20 – 1.80	1.70 – 0
				Observation conditions	4.00 – 3.60	3.50 – 3.00	2.90 – 2.40	2.30 – 0
				Completion and correctness of description of station	3.00 – 2.70	2.60 – 2.30	2.20 – 1.80	1.70 – 0
		Embedment	20	Marks and building materials	4.00 – 3.60	3.50 – 3.00	2.90 – 2.40	2.30 – 0
				Mark – stone types, specifications, sizes (including basic processing)	5.00 – 4.50	4.40 – 3.80	3.70 – 3.00	2.90 – 0
				Quality and stability of mark – stone casting	8.00 – 7.20	7.10 – 6.00	5.90 – 4.80	4.70 – 0
				Exterior decoration and warning	3.00 – 2.70	2.60 – 2.30	2.20 – 1.80	1.70 – 0

Table G. 2. 1 (Continued)

Branch quality	Score	Sub-item of quality	Score	Quality unit	Score			
					Excellent	Good	Qualified	Unqualified
Quality of observation achievements	60	Precision	25	The weakest point and RMSE of point position (total RMSE per kilometer)	12.50 – 11.30	11.20 – 9.40	9.30 – 7.50	7.40 – 0
				The weakest side and relative RMSE of side length (accidental RMSE per kilometer)	12.50 – 11.30	11.20 – 9.40	9.30 – 7.50	7.40 – 0
		Observation quality	25	Equipment inspection	3.00 – 2.70	2.60 – 2.30	2.20 – 1.80	1.70 – 0
				The rationality of observation method	3.00 – 2.70	2.60 – 2.30	2.20 – 1.80	1.71 – 0
				The standardization of observation	3.00 – 2.70	2.60 – 2.30	2.20 – 1.80	1.70 – 0
				Various tolerance conditions of observed values	6.00 – 5.40	5.30 – 4.50	4.40 – 3.60	3.50 – 0
				The standardization of the original record	4.00 – 3.60	3.50 – 3.00	2.90 – 2.40	2.30 – 0

Table G. 2. 1 (Continued)

Branch quality	Score	Sub-item of quality	Score	Quality unit	Score			
					Excellent	Good	Qualified	Unqualified
Quality of observation achievements	60	Observation quality	25	Implementation situation of design and specification	3.00 – 2.70	2.60 – 2.30	2.20 – 1.80	1.70 – 0
				The rationality of achievement re-survey and tradeoff	3.00 – 2.70	2.60 – 2.30	2.20 – 1.80	1.70 – 0
		Adjustment calculation	10	The progressiveness and legality of calculation program	3.00 – 2.70	2.60 – 2.30	2.20 – 1.80	1.70 – 0
				The rationality of the calculation achievement and adopted system	3.00 – 2.70	2.60 – 2.30	2.20 – 1.80	1.70 – 0
				The rationality and correctness of calculation plan, correction project and adopted data	4.00 – 3.60	3.50 – 3.00	2.90 – 2.40	2.30 – 0
Data compilation	10	Integrality of the data	7	Technical summary and report contents	4.00 – 3.60	3.50 – 3.00	2.90 – 2.40	2.30 – 0

Table G. 2. 1 (Continued)

Branch quality	Score	Sub-item of quality	Score	Quality unit	Score			
					Excellent	Good	Qualified	Unqualified
Data compilation	10	Integrality of the data	7	Acceptance records and documents of process inspection	1.00 – 0.90	0.89 – 0.75	0.74 – 0.60	0.59 – 0
				The submitted documents specified in design or contract	2.00 – 1.80	1.79 – 1.50	1.49 – 1.20	1.19 – 0
		Standardization of data	3	The standardization of data, such as description of station, original record, calculation results, etc.	1.00 – 0.90	0.89 – 0.75	0.74 – 0.60	0.59 – 0
				The standardization and print quality of technical summary report, inspection report, etc.	1.00 – 0.90	0.89 – 0.75	0.74 – 0.60	0.59 – 0
				The extent of meeting requirements for data compilation	1.00 – 0.90	0.89 – 0.75	0.74 – 0.60	0.59 – 0
Sub-item score of quality								

Note: This table shall be applicable to the control survey projects, such as GNSS survey, trigonometric survey, traverse survey, leveling, electro-optical distance measurement, special engineering surveying of river basin (area) control survey. If the project contents of a certain project contains only parts of the above contents, the 100 score system shall be adopted too, and the score shall be evaluated and endowed according to weight of sub-item or unit in respective segment.

Table G.2.2 The quality evaluation and scoring table of the digital terrain surveying and mapping achievement units

Branch quality	Score	Sub-item of quality	Score	Quality unit	Score			
					Excellent	Good	Qualified	Unqualified
The quality of digital surveying and mapping results	60	Control survey	30	Primary-level control	20	Rationality of points distribution and utilization conditions	5.00 – 4.50	4.49 – 3.75	3.74 – 3.00	2.99 – 0
					Quality of monumentation of mark-stones	5.00 – 4.50	4.49 – 3.75	3.74 – 3.00	2.99 – 0
					Point position accuracy	10.00 – 9.00	8.99 – 7.50	7.49 – 6.00	5.99 – 0
			Mapping control	10	Accuracy of mapping control survey	5.00 – 4.50	4.49 – 3.75	3.74 – 3.00	2.99 – 0
					Density of mapping control point	5.00 – 4.50	4.49 – 3.75	3.74 – 3.00	2.99 – 0
			Basis of figure	5	Correctness of the coordinates and elevation system	3.00 – 2.70	2.69 – 2.25	2.24 – 1.80	1.79 – 0
					Correctness of use of various projections and parameters	2.00 – 1.80	1.79 – 1.50	1.49 – 1.20	1.19 – 0

423

Table G. 2.2 (Continued)

Branch quality	Score	Sub-item of quality	Score	Quality unit	Score			
					Excellent	Good	Qualified	Unqualified
The quality of digital surveying and mapping results	60	Horizontal accuracy	15	Horizontal absolute or relative RMSE	12.00 – 10.80	10.79 – 9.00	8.99 – 7.20	7.19 – 0
				Edge matching precision	3.00 – 2.70	2.69 – 2.25	2.24 – 1.80	1.79 – 0
		Elevation accuracy	15	Height RMSE of the elevation points with note	6.00 – 5.40	5.39 – 4.50	4.49 – 3.60	3.59 – 0
				RMSE of contour lines	6.00 – 5.40	5.39 – 4.50	4.49 – 3.60	3.59 – 0
				Edge match accuracy	3.00 – 2.70	2.69 – 2.25	2.24 – 1.80	1.79 – 0
		Geographical accuracy	15	Rationality of comprehensive trade-offs	4.00 – 3.60	3.59 – 3.00	2.99 – 2.40	2.39 – 0
				Completion, correctness and coordination of geographical elements	5.00 – 4.50	4.49 – 3.75	3.74 – 3.00	2.99 – 0
				Correctness of the annotations and symbols	3.00 – 2.70	2.69 – 2.25	2.24 – 1.80	1.79 – 0

Table G. 2. 2 (Continued)

Branch quality	Score	Sub-item of quality	Score	Quality unit	Score			
					Excellent	Good	Qualified	Unqualified
The quality of digital surveying and mapping results	60	Geographical accuracy	15	Precision and rationality of geographical elements edge matching	3.00 – 2.70	2.69 – 2.25	2.24 – 1.80	1.79 – 0
		Correctness of data and structure	7	Rationality of the file name and data format	2.00 – 1.80	1.79 – 1.50	1.49 – 1.20	1.19 – 0
				Correctness and stratification of the factors layering	2.00 – 1.80	1.79 – 1.50	1.49 – 1.20	1.19 – 0
				Correctness and stratification of the attribute code	3.00 – 2.70	2.69 – 2.25	2.24 – 1.80	1.79 – 0
		Map decoration	3	Quality of symbols, lines, colors, annotations	1.00 – 0.90	0.89 – 0.75	0.74 – 0.60	0.59 – 0
				Coordination of graphic elements	1.00 – 0.90	0.89 – 0.75	0.74 – 0.60	0.59 – 0
				Map decoration, the marginal representation and framing	1.00 – 0.90	0.89 – 0.75	0.74 – 0.60	0.59 – 0

Table G. 2. 2 (Continued)

Branch quality	Score	Sub-item of quality	Score	Quality unit	Score			
					Excellent	Good	Qualified	Unqualified
Data compilation	10	Data integrity	7	Integrity of inspection report, technical summary report	4.00 – 3.60	3.59 – 3.00	2.99 – 2.40	2.39 – 0
				Process data	1.00 – 0.90	0.89 – 0.75	0.74 – 0.60	0.59 – 0
				Integrity of various of reports, maps, tables, atlas which specified by design and contract	2.00 – 1.80	1.79 – 1.50	1.49 – 1.20	1.19 – 0
		Standardization of data	3	Standardization of description of station, original record, calculation results, topographic map, etc.	1.00 – 0.90	0.89 – 0.75	0.74 – 0.60	0.59 – 0
				Standardization and print quality of technical summary report, inspection report, etc.	1.00 – 0.90	0.89 – 0.75	0.74 – 0.60	0.59 – 0
				Extent of meeting requirements for data compilation	1.00 – 0.90	0.89 – 0.75	0.74 – 0.60	0.59 – 0
Sub-item score of quality								

Note: The projects which are similar to underwater topography, water conveyance line and dike measurement of special engineering survey may refer to this table.

Table G. 2. 3 The quality evaluation and scoring table of water area survey (longitudinal – section survey, cross – sectional survey, underwater topographic survey)

Branch quality	Score	Sub – item of quality	Score	Quality unit	Score			
					Excellent	Good	Qualified	Unqualified
Quality of section survey achievements	60	Primary – level control	20	Rationality of points distribution and utilization conditions	5.00 – 4.50	4.49 – 3.75	3.74 – 3.00	2.99 – 0
				Quality of monumentation of mark – stones	5.00 – 4.50	4.49 – 3.75	3.74 – 3.00	2.99 – 0
				Point position accuracy	10.00 – 9.00	8.99 – 7.50	7.49 – 6.00	5.99 – 0
		Mapping control	10	Accuracy of mapping control survey	5.00 – 4.50	4.49 – 3.75	3.74 – 3.00	2.99 – 0
				Density of mapping control points	5.00 – 4.50	4.49 – 3.75	3.74 – 3.00	2.99 – 0
		Basis of figure	5	Correctness of the coordinates and elevation system	3.00 – 2.70	2.69 – 2.25	2.24 – 1.80	1.79 – 0
				Correctness of the use of mapping software and method	2.00 – 1.80	1.79 – 1.50	1.49 – 1.20	1.19 – 0

Control survey 30

Table G. 2. 3 (Continued)

Branch quality	Score	Sub-item of quality	Score	Quality unit	Score			
					Excellent	Good	Qualified	Unqualified
Quality of section survey achievements	60	Horizontal accuracy	15	Horizontal absolute or relative RMSE of center peg	12.00 – 10.80	10.79 – 9.00	8.99 – 7.20	7.19 – 0
				Distance tolerance from section point to section line	3.00 – 2.70	2.69 – 2.25	2.24 – 1.80	1.79 – 0
		Elevation accuracy	15	Height RMSE of center peg	7.00 – 6.30	6.29 – 5.25	5.24 – 4.20	4.19 – 0
				Error of section point height (difference)	8.00 – 7.20	7.19 – 6.00	5.99 – 4.80	4.79 – 0
		Geographic accuracy	15	Rationality of section points selection trade-offs and consistency between section and actual landform	6.00 – 5.40	5.39 – 4.50	4.49 – 3.60	3.59 – 0
				Integrality of contents and rationality of comprehensive trade-offs	6.00 – 5.40	5.39 – 4.50	4.49 – 3.60	3.59 – 0

Table G. 2. 3 (Continued)

Branch quality	Score	Sub-item of quality	Score	Quality unit	Score			
					Excellent	Good	Qualified	Unqualified
Quality of section survey achievements	60	Geographic accuracy	15	Correctness of annotations and symbols	3.00 – 2.70	2.69 – 2.25	2.24 – 1.80	1.79 – 0
		Correctness of data and structure	5	Correctness of file name and data organization	2.00 – 1.80	1.79 – 1.50	1.49 – 1.20	1.19 – 0
				Correctness of data format	2.00 – 1.80	1.79 – 1.50	1.49 – 1.20	1.19 – 0
				Rationality of mapping scale and method	3.00 – 2.70	2.69 – 2.25	2.24 – 1.80	1.79 – 0
		Map finishing	5	Quality of symbols, lines, colors, annotations	1.00 – 0.90	0.89 – 0.75	0.74 – 0.60	0.59 – 0
				Coordination of graphic elements	1.00 – 0.90	0.89 – 0.75	0.74 – 0.60	0.59 – 0
				Exterior finishing and framing of drawing and map – border, marginal representation and framing	1.00 – 0.90	0.89 – 0.75	0.74 – 0.60	0.59 – 0

Table G. 2. 3 (Continued)

Branch quality	Score	Sub-item of quality	Score	Quality unit	Score			
					Excellent	Good	Qualified	Unqualified
Data compilation	10	Data integrity	7	Integrity of inspection report, technical summary report	4.00 – 3.60	3.59 – 3.00	2.99 – 2.40	2.39 – 0
				Process data	1.00 – 0.90	0.89 – 0.75	0.74 – 0.60	0.59 – 0
				Integrity of various of reports, maps, tables, atlas which specified by design and contract	2.00 – 1.80	1.79 – 1.50	1.49 – 1.20	1.19 – 0
		Standardization of data	3	Standardization of description of station, original record, calculation results, section diagram, etc.	1.00 – 0.90	0.89 – 0.75	0.74 – 0.60	0.59 – 0
				Standardization or print quality of technical summary report, inspection report, etc.	1.00 – 0.90	0.89 – 0.75	0.74 – 0.60	0.59 – 0
				Extent of meeting requirements for data compilation	1.00 – 0.90	0.89 – 0.75	0.74 – 0.60	0.59 – 0
Sub-item score of quality								

Note: The projects which are similar to water conveyance line, dike measurement and geological survey of special engineering survey may refer to the table.

Table G. 2. 4 The quality evaluation and scoring table of boundary markers survey and layout in resettlement engineering survey

Branch quality	Score	Sub-item of quality	Score	Quality unit	Score			
					Excellent	Good	Qualified	Unqualified
Quality of boundary marker achievements	60	First-level control	20	Rationality of points distribution and utilization conditions	5.00 – 4.50	4.49 – 3.75	3.74 – 3.00	2.99 – 0
				Quality of monumentation of mark – stones	5.00 – 4.50	4.49 – 3.75	3.74 – 3.00	2.99 – 0
				Point position accuracy	10.00 – 9.00	8.99 – 7.50	7.49 – 6.00	5.99 – 0
		Mapping control	10	Accuracy of mapping control survey	5.00 – 4.50	4.49 – 3.75	3.74 – 3.00	2.99 – 0
				Density of mapping control point	5.00 – 4.50	4.49 – 3.75	3.74 – 3.00	2.99 – 0
		Survey method	10	Rationality of the survey plan	5.00 – 4.50	4.49 – 3.75	3.74 – 3.00	2.99 – 0
				Correctness of determination of the accuracy grade	5.00 – 4.50	4.49 – 3.75	3.74 – 3.00	2.99 – 0

Table G. 2. 4 (Continued)

Branch quality	Score	Sub-item of quality	Score	Quality unit	Score			
					Excellent	Good	Qualified	Unqualified
Quality of boundary marker achievements	60	Horizontal accuracy	10	Horizontal absolute or relative RMSE	4.00 – 3.60	3.59 – 3.00	2.99 – 2.40	2.39 – 0
				Conform to survey achievements used in design phase	3.00 – 2.70	2.69 – 2.25	2.24 – 1.80	1.79 – 0
				Coordinate difference between design phase and completion phase	3.00 – 2.70	2.69 – 2.25	2.24 – 1.80	1.79 – 0
		Elevation accuracy	20	RMSE of height	6.00 – 5.40	5.39 – 4.50	4.49 – 3.60	3.59 – 0
				Confirm with the survey achievements used in design phase	4.00 – 3.60	3.59 – 3.00	2.99 – 2.40	2.39 – 0
				Elevation difference between design phase and completion phase	10.00 – 9.00	8.99 – 7.50	7.49 – 6.00	5.99 – 0

Table G. 2. 4 (Continued)

Branch quality	Score	Sub-item of quality	Score	Quality unit	Score			
					Excellent	Good	Qualified	Unqualified
Quality of boundary marker achievements	60	Quality of boundary markers manufacture and monumentation	15	Monumentation specification and production quality	5.00 – 4.50	4.49 – 3.75	3.74 – 3.00	2.99 – 0
				Stability of the monumentation	5.00 – 4.50	4.49 – 3.75	3.74 – 3.00	2.99 – 0
				Depth and size outcropped of ground of monumentation	5.00 – 4.50	4.49 – 3.75	3.74 – 3.00	2.99 – 0
		Drawing of the boundary marker points and water level	5	Correctness of reallocated population, resettlement and land aquisation	2.00 – 1.80	1.79 – 1.50	1.49 – 1.20	1.19 – 0
				Correctness of the inundated line	1.00 – 0.90	0.89 – 0.75	0.74 – 0.60	0.59 – 0
				Quality of symbols, lines, color, note	2.00 – 1.80	1.79 – 1.50	1.49 – 1.20	1.19 – 0
Data compilation	10	Data integrity	7	Integrity of inspection report, technical summary report	4.00 – 3.60	3.59 – 3.00	2.99 – 2.40	2.39 – 0
				Process inspection data	1.00 – 0.90	0.89 – 0.75	0.74 – 0.60	0.59 – 0

Table G. 2. 4 (Continued)

Branch quality	Score	Sub-item of quality	Score	Quality unit	Score			
					Excellent	Good	Qualified	Unqualified
Data compilation	10	Data integrity	7	Integrity of various of reports, diagrams, atlas, trusteeship book, outcome table, etc. which specified by design and contract	2.00 – 1.8	1.79 – 1.50	1.49 – 1.20	1.19 – 0
		Standardization of data	3	Standardization of description of station, original record, calculation achievements, trusteeship book, distribution diagram of point positions, outcome table, etc.	1.00 – 0.90	0.89 – 0.75	0.74 – 0.60	0.59 – 0
				Standardization and print quality of technical summary report, inspection report, etc.	1.00 – 0.90	0.89 – 0.75	0.74 – 0.60	0.59 – 0
				Extent of meeting requirements for data compilation	1.00 – 0.90	0.89 – 0.75	0.74 – 0.60	0.59 – 0
Sub-item score of quality								

Note: The control survey and topographic survey of immigrant engineering survey in special engineering survey may refer to Table G. 2. 1 – Table G. 2. 4.

Table G. 2. 5 The quality evaluation and scoring table of results data editing and warehousing

Branch quality	Score	Sub-item of quality	Score			
			Excellent	Good	Qualified	Unqualified
Quality of importing data into database	20	Integrity of data storage file	5.00 – 4.50	4.49 – 3.75	3.74 – 3.00	2.99 – 0
		Integrity of the data	5.00 – 4.50	4.49 – 3.75	3.74 – 3.00	2.99 – 0
		Standardization of data format and exchange format	5.00 – 4.50	4.49 – 3.75	3.74 – 3.00	2.99 – 0
		Meeting requirements of database, technical specification and design	5.00 – 4.50	4.49 – 3.75	3.74 – 3.00	2.99 – 0
Established database quality	30	Progressiveness, openness and practicability	10.00 – 9.00	8.99 – 7.50	7.49 – 6.00	5.99 – 0
		Standardization, formalization and versatility	10.00 – 9.00	8.99 – 7.50	7.49 – 6.00	5.99 – 0
		Networked function	10.00 – 9.00	8.99 – 7.50	7.49 – 6.00	5.99 – 0
Coordinated system	10	Consistency between spatial reference coordinates and the database	2.00 – 1.80	1.79 – 1.50	1.49 – 1.20	1.19 – 0
		Unity of coordinate systems	2.00 – 1.80	1.79 – 1.50	1.49 – 1.20	1.19 – 0
		Conformance between the horizontal and vertical system and national standards	3.00 – 2.70	2.69 – 2.25	2.24 – 1.80	1.79 – 0
		Conversion relations between independent coordinate system and national unified system	2.00 – 1.80	1.79 – 1.50	1.49 – 1.20	1.19 – 0
Data dictionary	15	Normalization	5.00 – 4.50	4.49 – 3.75	3.74 – 3.00	2.99 – 0
		Applicability	5.00 – 4.50	4.49 – 3.75	3.74 – 3.00	2.99 – 0
		Clearness and integrity	5.00 – 4.50	4.49 – 3.75	3.74 – 3.00	2.99 – 0
Operation of database	25	Security	5.00 – 4.50	4.49 – 3.75	3.74 – 3.00	2.99 – 0
		Stability	5.00 – 4.50	4.49 – 3.75	3.74 – 3.00	2.99 – 0
		Maneuverability	5.00 – 4.50	4.49 – 3.75	3.74 – 3.00	2.99 – 0
Sub-item score of quality		Inspection result of professional legal organization	10.00 – 9.00	8.99 – 7.50	7.49 – 6.00	5.99 – 0

G. 3 Compilation of Surveying and Mapping Product Inspection Report

G. 3. 1 The process inspection of surveying and mapping products does not undergo the product quality grade evaluation, and the process inspection report of surveying and mapping products may not be compiled, but the process inspection table shall be filled in.

G. 3. 2 The format of surveying and mapping process inspection tables, shall be implemented in accordance with Table G. 3. 2.

Table G. 3. 2 The format of the surveying and mapping process inspection table

Surveying and mapping process inspection table						
						No.
Project name			Design phase			
Product name			Executing units			
Main contents of product:							
No.	Problems record		Treatment opinions	Treatment		Recheck Situation

G. 3. 3 The inspection report of surveying and mapping products shall be implemented according to the following stipulations:

   1 As for the covering of the surveying and mapping product inspection report, see Table G. 3. 3 - 1.

Table G. 3. 3 - 1 The covering format of the surveying and mapping product inspection report

Serial No:
The inspection report of surveying and mapping products
Project name: Product name: Manufacturer: Inspection department:
Date: _____

2 The main body of the surveying and mapping product inspection report shall be implemented according to Table G. 3. 3 – 2.

Table G. 3. 3 – 2 The main body format of the surveying and mapping product inspection report

\multicolumn{4}{c}{The inspection report of surveying and mapping products}
		The 1 page	Total page
Project name		Design phase	
Product name		Executing units	
Report writer (sign):		Y M D	
Main contents of product:			
Conclusions of inspection departments:		Quality grade	
	Position:	Signature: Y M D	
Opinions of technical director:			
	Position:	Signature: Y M D	
Opinions of executive leadership:			
	Position:	Signature: Y M D	
Remarks:			

Table G. 3. 3 - 2 (Continued)

Page 2, pages in total

1. Task summary:
2. Inspection condition (including instruments and personals):
3. Technical basis of inspection:
4. Main contents of inspection:
5. Main quality problems and amend opinions:
6. Amend opinions of leftover problem:
7. Inspect conclusion:
Writer (sign):   Date: _____

3 As for the format of the attached table of the surveying and mapping product inspection report, see Table G. 3. 3 – 3.

Table G. 3. 3 – 3 The format of the attached table of the surveying and mapping product inspection report

The opinion record table of inspection report					
Page, pages in total					
Project name			Design phase		
Product name			Manufacturer		
Inspection parameter: project plan (Data quality/Position quality/Material quality/...)					
No.	Quality problems	Treatment opinions	Amend condition	Review condition	Error category
Remarks:					
The final inspectors: Date: The reviewer: Date:					